Word Problems for Smart Kids

Keep your Child Trained with Intriguing Word Problems

D1799185

Brendon Stock

prohibited and any storage of this document is not allowed unless with written permission from the publisher. All rights reserved.

The information provided herein is stated to be truthful and consistent, in that any liability, in terms of inattention or otherwise, by any usage or abuse of any policies, processes, or directions contained within is the solitary and utter responsibility of the recipient reader. Under no circumstances will any legal responsibility or blame be held against the publisher for any reparation, damages, or monetary loss due to the information herein, either directly or indirectly.

Respective authors own all copyrights not held by the publisher.

The information herein is offered for informational purposes solely, and is universal as so. The presentation of the information is without contract or any type of guarantee assurance.

The trademarks that are used are without any consent, and the publication of the trademark is without permission or backing by the trademark owner. All trademarks and brands within this book are for clarifying purposes only and are the owned by the owners themselves, not affiliated with this document.

INTRODUCTION

The Word Problems are listed by grade and by theme within each grade. I still find it helps to keep my daughter excited about doing her job by having a seasonal worksheet.

The grade levels are a guideline — please use your discretion based on your child's skill and eagerness (my older daughter has always used a grade below, while my younger daughter appears to be a grade or two above — go figure). Bear in mind that Math Word issues require reading, comprehension, and math skills, so when faced with math word problems, a good at simple math equations will struggle more than you would think.

All word issues are dynamic. There will be no shift in the words in a specific problem, but the numbers will.

Children struggling to turn a word problem into a math equation would find it comforting to revisit the same verbal hints with different numbers (confidence builder), so consider printing a few regenerations of each problem. You may provide partners or a group of students with a problem to solve together in a classroom environment and then provide regeneration of the same problem for the children to do solo.

With my eldest daughter, I sometimes walked through a math problem with her (doing much of the work myself) until I realized how much she struggled with math when it was not written down in a nice tidy equation. I then gave her a few regenerations with distinct numbers of the same problem for her to do solo. After a few weeks of this, without Mom's walkthrough, she was able to do them. She's one of those children who say, "It's too hard!" very easily, so it's important to create trust — if she thinks she can't do anything she can't do — if she thinks she can do anything she can do. How can I persuade her now that she can keep her room clean?

CHAPTER 1: WORD PROBLEMS

Word problems are among the first ways we see math implemented and one of the most anxiety-producing math difficulties faced by many grade school children. This section has a great set of word issues for all four basic math operations that include a gentle introduction to word problems. You'll find addition word issues, word subtraction issues, word multiplication issues, and word division issues, all beginning with basic easy-to-solve issues that work up to the more nuanced skills needed for several standardized tests. You can also encounter a mix of activities as they progress, requiring students to figure out what kind of story issue they need to address. And at the bottom of this list, check out word problem tricks if you need help!

The awful bugaboo, a math word dilemma, is one of the greatest challenges your youngster has to conquer in school. The one criticism I've encountered all too many times during my many years of private instruction is the inability to solve the word issue. Yet word issues can be successfully addressed. This paper clarifies how.

Word problems are more difficult than "ordinary" math problems because you have to determine what needs to be achieved and how to do it, and you need one solution. Thus, unlike the solution of an equation such as $x + 3 = 4$, and then asking for the value of x, a word problem allows one to decide from the words what equations can be derived and how to solve those particular equations.

Another obstacle lies in the student's inability to read at a level sufficient to make sense of the terms that make up the problem. Generally, poor readers can produce poor word problem solvers. This is why I teach essential reading skills to students, including methods such as "anticipatory reading" and other active competencies in reading. These strategies provide students with tremendous improvement in their mathematical skills and transfer to other disciplines, including reading, social studies, and English.

To better understand these methods, at the pre-algebra / algebra stage, we will look at a particular word problem and then see how to apply those techniques. The subject we're going to address is the issue of algebra equation systems.

Word Problem Example: It costs $23 for five hockey sticks and three hockey pucks. It costs $20 for five hockey sticks and one hockey puck. How much are two pucks going to cost?

Strategies for Word Problems:

First Pass: This is the stage where we just read the issue to get a "feel" for what's going on. We do not try to solve the real issue during this process but get an overall impression.

Second Pass: This is the stage where we re-read the topic, paying close attention to the situation at hand, what the issue is about, who are the key players, and so on. During this point, we begin to mull over some problem-solving tactics and begin to prepare our attack.

Third Pass: This is the brainstorming stage. We decide the essence of the issue at this stage, what we know, and what we are asked to do. This is when we begin to translate words and calculate all inside the problem to numbers and equations.

Fourth Pass: This is when we use the knowledge we obtained in the third pass to solve the problem. We also double-check our brainstorming process at this stage to ensure we have taken the correct approach.

Fifth Pass: This is the final stage in which we evaluate the fourth pass's solution for accuracy.

With the topic at hand, let us go through these phases. In the first pass, we read the issue and see that it has everything to do with hockey sticks and hockey pucks and the cost of two pucks. Note that a curveball has been thrown here in that we are asked to state the price of two pucks, not one. Keep this in mind for the issue to stop.

Now, during the second pass, we find that we're only dealing with hockey sports, that we're limited to two pieces of equipment, pucks, and sticks, and that we're given prices for some of the two variations, and that we're explicitly asked for the price of two pucks.

We start constructing the initial mathematics on the third pass. We got five sticks, and they cost $23 for three pucks. We also realize it costs $20 for 5 sticks and 1 puck. We can even guess some numbers that might work to make sure we have a clear understanding of the problem. You might guess, for instance, that a stick might be $4, and a puck would be $1. It will then cost $23 for 5 sticks and 3 pucks, which seems like a fair option. However, the second condition does not satisfy certain values: 5 sticks and 1 puck cost $20. Know, to be the right ones, and the final values have to satisfy all conditions. But at least we are, with our initial guess, in the ballpark.

We pick letters to describe our things in the problem in our fourth pass, and then we bring our equations together. Given that we are concerned with pucks and sticks, S for stick and P for the puck would be a reasonable choice of letters. Gee. Gee. Gee. Really? Really? All right, so we've got the following two equations now:

1. $5S + 3P = \$23$

2. $5S + 1P = \$20$

Now you see that a simple system of linear equations is being looked at. You can solve it by using the elimination process. Therefore, if we subtract equation 2 from equation 1, we end up with $2P = \$3$ or, by simple division, $P = \$1.50$. If we plug this value back into equation 1 for P, we get S = $3.70. We have $2 \times \$1.50 = \$3.00$ now, going back to what was asked for, the price of two pucks.

We should ask ourselves if, in the fifth step, our answer is fair. It seems like the stick's cost should be higher than the cost of the puck, even though the price of the stick seems a little cheap. We get a check if we plug these values for S and P into equation 2, and we can therefore feel assured that our solution is correct.

Your children can confidently solve word problems using this basic phase technique, whether the problem involves hockey pucks and sticks, giraffes and elephants, or whether the solution involves equation systems or mixed rate problems. In the often called ghoulish world of word problems, critical reading, effectively solving, and applying this five-step method will ensure impressive success. Beware goblins!

CHAPTER 2: WORD PROBLEMS FOR INCLUSION

Besides, these introductory word problems are perfect for the first or second grade applied mathematics.

Subtraction Problems with Terms

20 Worksheets on Word Problems

These worksheets contain basic word problems with smaller amounts for subtraction. Watch for phrases like gap and rest.

Term Mixed Addition and Subtraction Issues

This worksheet collection contains a combination of word problems with addition and subtraction. Students are needed, provided the problem context, to find out which operation to apply.

Term Multiplication

This is the first set of worksheets for word problems that multiplication implements.

Word Division Issues

These division story issues deal only with whole divisions (quotients without residues). This is a fantastic first step in understanding the keywords that signify a problem's division word problem.

Cookie Division Girl Scout

If you've served as a Troop Cookie Mom (or Daddy!), you'll know what kind of math we've been doing ... These worksheets are mostly issues with division terms that introduce remainders. Take out the box of your tagalongs or your thin mints and find out how many reminders you should eat!

Word Issues Division Of Remainders

In this segment, the worksheets are made up of story problems using division and involving residues. These are equivalent, but with different units, to the Girl Scout issues in the previous section.

Mixed Word problems of multiplication and division

These worksheets incorporate the problems of simple multiplication and word division. The division concerns do not include leftovers. These worksheets enable students to distinguish between the phrasing of a story problem requiring multiplication versus one requiring the answer to be reached by the division.

We can learn how to solve the word problems of multiplication and division of whole numbers step-by-step. In our everyday lives, we know we need to do multiplication and division. Let us solve some examples of a word query.

Word problems with big numbers of multiplication and splitting:

1. The cost of a chair is 980.50 dollars. Find the expense of the 2035 chairs like that.

Solution:

Cost of a chair =	980.50
Number of chairs =	× 2035
	490250
	2991500
	0000000
	+ 196100000
Cost of 2035 chairs =	1995317.50

Therefore, cost of 2035 chairs is $1995317.50.

2. A tyre factory produces 6348 tyres a day. How many tyres will the factory produce in 460 days?

Solution:

Number of tyres a tyre factory produces in one day
=   6348

Number of day =
× 460

0000

380880

                                                    +

2539200

Total number of tyres produces in 460 days =
2920080

Hence, in 460 days, the tyre factory will
manufacture 2920080 tyres.

3. The cost is $4218000 for a flat. If a building has
36 identical apartments, how much money is raised
by selling all the flats?

The Solution:

Cost of one flat = $4218000

Number of flats of similar nature = 36

Cost of 36 flats = $4218000 × 36

Hence, the money collected = $151848000

4. 470988 books on shelves should be arranged fairly. How many shelves are needed if 378 books are placed on each shelf?

Solution:

Total number of books = 470988

Number of books arranged on each shelf = 378

Number of shelves needed = 470988 ÷ 378

Therefore, 1246 shelves needed to arrange 470988 books equally.

5. The cost was $457104 for 534 train tickets. Check the expense of a single ticket?

The Solution:

Number of tickets for the train = 534

Cost of 534 train tickets = $457104

Cost of 1 ticket by train = 457104 ÷ 534

The cost of one train ticket, therefore, is $856.

6. Thirty-six families went on a trip that cost them $1216,152. If each family fairly shared the expense, how much did each family pay?

The Solution:

The complete number of households = 36

Total cost paid by 36 families = $1216152 Total cost paid by 36 families

Cost paid by each family = $1216152 ÷ 36

Hence, the cost paid by each family = $33782

Problems with Word Mixed Operation

A whole enchilada! These worksheets incorporate word problems with addition, subtraction, multiplication, and division. These worksheets will test a student's ability to select the correct operation based on the problem text of the plot.

Addition of Word Problems with Additional Facts

Using extra (but unused) information in the problem text is one way to make a word problem slightly more complicated. These worksheets have additional word problems with the issue of additional unused facts.

Subtraction Word Issues Additional facts.

For subtraction, Word problem worksheets with extra unused facts in each question. The worksheets begin with problems of subtraction with smaller values and progress through more complicated issues.

Extra Facts Term issues of addition and subtraction

Mixed operation addition and word problem subtraction worksheets with extra unused information in the issues.

Multiplication issues with Term Extra Facts

Word multiplication problems with extra unused facts in the problem. The worksheets begin with multiplication problems with smaller values and advance through more difficult issues in this package.

Extra Facts Division Issues with Term

Math word issues for division with extra unused facts in the problem are included in this section's worksheets. No residues are found in the quotients in these division problems.

Multiplication and division issues of Term Extra Facts

This is a set of worksheets with mixed multiplication and word division problems and additional unused facts in the problem. No residues are found in the quotients in these division problems.

Word Issues in Travel Time (Customary)

These story issues deal with travel time using miles (customary units) to calculate the travel distance, travel time, and speed. This is a very popular word problem class, and specific experience with these worksheets will prepare students for standardized tests when they encounter similar issues.

SAT Math: Score High by Knowing How to Translate From English Into Algebraic Notation,

Part I: Translations Involving Only One Arithmetic Operation Students who score high in the SAT Math are not necessarily better at computation or equation solving. They are better at is "problem translation," the ability to translate English expressions into algebraic notation. The good news about problem translation is that little more is required than memorizing a set of rules. In the same way that, when taking a foreign language, you memorize the fact that "house" is "Haus" in German, "maison" in French, or "casa" in Spanish, you memorize the fact that "three less than a number" is written as "x - 3" in algebraic notation.

Words indicating addition are plus, sum, total, greater than, increase, increased by, more than. Thus, "the sum of a number and 7" would be written as "n + 7." Notice that we can use any letter as a variable-x, n, w, or w or any other letter we choose.

In addition to recognizing the words that indicate addition, recognize the situations in which addition is logically implied. Generally, the idea of increase will imply addition.

Subtraction is a little trickier because often -- but not always -- the word order of an English phrase indicating subtraction will reverse the order of the algebraic expression. Words indicating subtraction are minus, difference, and less.

Here are some common phrases indicating subtraction and how the corresponding algebraic expressions are written: A number minus three n-3 The difference between a number minus three n-3 The difference between a number minus three n-3 The difference between a number minus three n-3 The difference between a number and 5 x-5 3 less than a number x-3 (Notice the reverse of the order.) 7 is subtracted from a number x - 7 (Notice that the word "from" indicates a starting point. For example, if you say that you traveled to New York from Cleveland, that phrase indicates that you started from Cleveland. In algebra, you "started" from x. Hence, the x should come first in this expression.

When translating into algebra, even if the keywords are not there, the problem implies that a decrease has taken place, subtract. For example, if someone talks about losing money or the stock market going down, a decrease occurs. Usually, that will indicate subtraction.

Multiplication:

1.   Look for words like products and times.

2.   More importantly, look at the underlying logic of the problem. If the same thing is repeatedly happening and you are trying to find a total amount, think about multiplying.

3.   Recognize that the word "of" when accompanying a fraction or decimal indicates multiplication.

For example, if a landlord owns ten apartments and collects $800 from each tenant, his total income is 10(800) = $8,000.

Division:

1.   Look for words like divide and ratio.

2.   More importantly, look for situations in which the total is known, but each part's size is not known.

3.   Suppose we reverse the landlord's information, and we know that he has 10 apartments and earns $8,000 per month.

If we assume that each apartment rents for the same amount, how much is the rent for each apartment? In this case, we know a total amount and are trying to find the size of each equal part-in in this case, a rent payment. Therefore, we divide.

Important Point: Distinguish between the meanings of "greater than," "is greater than," and "how much greater than." Greater than indicates addition. 8 greater than x x + 8 Is greater than indicates an inequality 8 is greater than 5 8 > 5 How much greater than indicates subtraction. "How much greater than 5 is x" would be written as

x -- 5. Similarly, distinguish between "less than," "is less than," and "how much less."

Less than indicates subtraction. 7 less than a number is x - 7.

It is less than indicates inequality. 3 is less than 6 is 3 < 6.

How much less than indicates subtraction How much less than 8 is x 8 - x

Practice the following Translation Exercises: Use x to represent an unknown number.

1. A number increased by 11.

2. The sum of a number and 13

3. 3/4 of a number

4.. A number minus 17

5. 12 less than a number.

6. 15 is subtracted from a number

7. The ratio of a number and 20.

8. The number of milk cartons is divided evenly among 15 children.

9.. How much greater is 59 than a number.

10. Total profit is x items are sold and earn $6 profit each.

Answers:

1. x + 11

2. x + 13

3. 3/4(x) (Three-fourths times x)

4. x - 17

5. x -- 12

6. x -- 15

7. x / 20 (x divided by 20.)

8. x/ 15 (x divided by 15)

9. 59 -- x

10. 6x

Travel Time Word Problems (Metric)

Did I wonder when the train will arrive? These story problems deal with travel time using kilometers (metric units) to calculate travel time, travel time, and speed.

Taking the words into a workable mathematical equation, and translating them is the hardest thing about doing word problems. Many students fear and hate doing it for this purpose. Knowing where to begin and how to find out the solution may not be very obvious. There are, however, ways to break down a dilemma with a term that makes it simpler and easier to solve. The following is a list of helpful vocabulary tips and strategies to overcome these tough problems.

1. Know that you are trying to turn the words into an equation when doing a word problem, so read through the whole problem first. Don't try to solve the problem when you've just read one sentence. It is important to read the problem thoroughly to get

the whole picture and effectively interpret and solve it.

2. Go back to the start. Only re-read the first line. Write down what is known and what is unknown to you. Use variables to represent the unknowns and mark what they stand for clearly. Do the same for the second sentence and for each sentence that follows.

3. Look for keywords that indicate certain mathematical operations when you are doing this:

ADDITION

Increased by more than the plus total combined added to combined

SUBTRACTION

Reduced by the difference of less than how many more? More than minus (from) subtraction

MULTIPLICATION

Multiplied by twice by-product of times

DIVISION

A share ratio separated by an equivalent proportion of the percentage quotient

EQUALS

The gifts are sold so that the yields will be

4. Perform the operations necessary.

Let's use these approaches as an example:

456 students attend Lincoln Middle School. One-third of the students take part in athletics. This is two-fifths of the number of students in a high school nearby. How many high school students are attending?

MS of Lincoln = 456 students

$1/3(456)$ = students involved in sports = 152

x = number of HS students

$152 = 2/5 \, x$

$5/2 \, (152) = 5/2 \, (2/5)x$

$380 = x$

The high school has 380 students.

5. Make sure the question that is being asked of you is answered. To arrive at the final answer, it often takes several steps, and you may lose sight of what the word problem asks for or asks you to do. The response may seem to be one step, such as the

second step above 1/3(456) = sports students = 152. It might seem like the number 152 is the solution, but it's just the number required to solve the problem. Remember to include any measurement units required in your response, such as ft., cm, lb, oz, etc. Label the response with the measurement unit or with the item that the query asks. "For instance," 380 students from high school.

6. Finally, check your answer mentally or on paper to make sure it is correct and makes sense. You do this by replacing the variable with your response in each step and calculating the equation with the variable known to see if it satisfies the equation. If it does, then you have finished!

To master word issues, note, and you need to practice, practice, practice.

Many children struggle from the very beginning with mathematics. Since math is a progressive study focused on early fundamentals, building one definition on top of another requires a solid understanding of the subject. For this reason, every school year, math gets harder for these children.

Rate Word Problems: Speed, Distance, and Time

Distance, rate, and time problems are standard applications of linear equations. When solving these problems, use the relationship **rate** (speed or velocity) times; time equals **distance**.

For example, suppose a person were to travel 30 km/h for four h. To find the total distance, multiply rate times or (30km/h)(4h) = 120 km.

The problems to be solved here will have a few more steps than described above. So to keep the information in the problem organized, use a table.

Distance, time, and rate problems have a few variations that mix the unknowns between distance, rate, and time. They generally involve solving a problem that uses the combined distance traveled to equal some distance or a problem in which the distances traveled by both parties are the same. These distance, rate, and time problems will be revisited later on in this textbook, where quadratic solutions are required to solve them.

CHAPTER 3: TRICKS TO OVERCOME WORD ISSUES

On this portion of the section, the math worksheets deal with basic word issues suitable for primary grades. Depending on student ability, the basic addition word issues may be implemented very early in first or second grade. If the subtraction principle is addressed, follow these worksheets with the subtraction word problems, and then continue in the same way with multiplication and division word problems.

For students, word issues are also a source of anxiety because we prefer to incorporate math operations in the abstract. Students fail to relate even elementary operations to word problems if they have been trained regularly to think about math operations in their day-to-day routines. When asked frequently, talking with children about 'how many more do you need' or 'how many do you have left over' or other seemingly simple questions will establish the basic sense of number that helps immensely when word problems and applied math begin to show up.

There are some tricks to overcome word issues that can bridge the gap, and they can be useful tools if students are either struggling with where to start with a problem or need a way to check their thoughts on a particular topic.

Make sure that your student first reads the entire issue. It's really simple to start reading a word problem and think, 'I know what they're asking for ...' after the first sentence or two and then have the problem take an entirely different turn. It can be complicated to overcome this early solution bias. Before deciding on a route to the solution, it is much easier to cultivate the habit of making a full pass over the problem.

For various operations, unique words tend to turn up in word problems, which could tip you off to the correct operation to submit. Such keywords are not a sure-fire way to understand what to do with a problem, but they can be a helpful starting point.

For instance, phrases such as 'combined," complete," together' or 'amount' often suggest that addition would involve the problem.

Subtraction word problems in their wording often use 'difference," less,' or 'decrease.' Word problems will also use verbs such as 'gave' or 'shared' as a stand-in for subtraction for younger children.

Clear ones such as 'days' and 'product,' and be on the lookout for 'for each' and 'every' is the key phrases to watch out for multiplication word issues.

It can be tricky to learn when to apply division in a word problem, particularly for younger children who have not completely developed a concept of what division can be used for ... But that's exactly why split word issues can be so beneficial! It should sound loud and clear to your division radar if you see words like 'per' or 'among' in the word problem text. Pay attention to 'shared among' and ensure that this phrasing is not confused with a subtraction word problem by students. That is a good example of when it is really important to pay attention to the language.

Draw a picture!

One main piece of advice is to encourage students to draw an image, particularly for simple word problems. Simple counting activities, where you work with amounts or sets that are fairly small, are most early grade school word issues. If students can draw an image of the problem (even using basic representations for the units discussed in the problem, such as squares or circles), it can allow them to imagine exactly what is happening.

The use of manipulatives is another effective technique for visualization. Paper clips, checkers, or other useful items may stand in place of the subject of the problem. This offers an opportunity to work with different numbers on other basic examples.

GRASS: GRASS

Provided: Describe in the query the information given. What values, and what do they represent, are given? Knowing the following words that are often used in math questions and what they mean can be helpful:

Term Meaning

Sum + (addition)

Difference - (subtraction)

Product * (multiplication)

Quotient / (division)

Required: Look at what you want to find in the question. Represent it with a variable, whatever it is, e.g., x.

ANALYSis:

If applicable, if one is not already provided, sketch.

Decide on the appropriate mathematical methods to be used and the formulas relating to various interest quantities.

Use one variable only, whenever possible.

For instance, if you're looking for two numbers that differ by 5, let the two numbers, NOT x and y, represent x and x+5 (or x and x-5). Alternatively, to help you rewrite your equation in terms of one variable only, you may be able to use some information presented in the question.

Solution: Solve or find the answer you're looking for in the equation. To figure out which ones work for this specific issue, you will have to think hard about the different math techniques you have learned in your courses. In the context of the application, ensure that your solution(s) makes sense, e.g., a negative area makes no sense.

Statement: To make clear the answer to the question, finish with a concluding statement. Include units in your answer when applicable.

To demonstrate how GRASS is used to solve a word problem here is a simple example.

EXAMPLE:

The perimeter of a park with rectangles is 26 m. What is the width of the park if the length of the park is 5 m?

Given: The perimeter of the rectangular park is given as 26 m in this question. We are also told that the park's length is 5 m. So, P= 26 m and L= 5 m, respectively.

Required: The question is: "What is the park width?" We know from this that we must find the width. Locate: w =?

Analysis: To determine what formulas we might have to use to find the width, we can sketch the rectangular park.

Note: the lengths are the longer sides, and the widths are the shorter sides.

We can see from this sketch that we need a formula for the perimeter of a rectangle.

P= 2L + 2w 2W +

If we rearrange the equation for w, we get the equation for w.

W = (P-2L) ÷ 2 2

The streamlined equation is,

w = P/2-L

Solution: So, considering that we have our equation, the unknown variable w can be set by subbing our known variables, P and L.

w = (26m)/2 - 5m

w = 13m - 5m

w = 8m

Statement: We have found the width to be 8 meters. So now, write down your final answer in words with your solution.

Therefore, 8 meters is the width of the rectangular park.

Support Overcome Word Problems

You've come to the right place if you need help solving word problems. Because of how logical and accurate it is, math has always been a fascinating subject for me. It's close to learning another language and very much like solving a puzzle, in the case of solving word problems. You have to be a good detective and pick up on clues that will help you solve the mystery if you become good at solving word problems. Working with word issues includes knowledge of reading, as well as the capacity to solve math equations. With that said, this piece aims to offer support to solve word problems and provide a strategy for solving these problems for aspiring mathematicians.

Let's start with the simple example below:

The length of a basketball court in the NBA is 44 feet longer than its width. Express the court's length in terms of its width.

Simple but critical is the first step to dealing with word issues. Read the entire issue, please ... then read it again! This is where reading comprehension skills come into play, as simple as this might sound. It is during this critical period that the following must be done:

1. Identify the knowledge you possess:

The court is about 44 feet wider than the wide one.

2. Identify the data you don't (and do need) have:

The court's length.

3. Determine what is called for by the word problem:

In terms of distance, an equation expresses length.

Starting to arrange your clues is the next move. Begin by assigning variable names first to pieces of information that you have and don't have. Clear and relevant should be the name(s). So we'll allocate the variable name 'L' and the width of the variable name 'W' to the length of the basketball court.

Drawing them out on paper is one thing that has helped me grasp concepts. Usually, when a concept can be visualized, it is easier for them to grasp it, whether it is mathematics or something else. So, draw an image after you define the data that you have and assign your unknown variables. Be sure to mark it with known info and variables that are unknown.

The last step is looking for keywords in the word problem. In seeking the solution, certain terms will tell you what mathematical activity is required or at work. There are a few of these words mentioned in the table below:

Addition: Absolute, number, plus, more than, all / together, combined, all, or plus.

The difference, less / less than, reduced by, minus, or less. Subtraction:

Multiplication: multiplied by, by time, by, or by-product.

Division: the ratio of, quotient of, or percent, per, a, out of.

Equals: is, is, was, has been, gives, or yields.

Using our example of a basketball court, we are told that the court's length is 44 feet longer than its width. "In this sentence, there are two keywords / clues that we can identify;" more than, "which means that the mathematical operation that will be at work here will be added (+) and" is, "which can be interpreted as" equals "or" =.

To come up with an equation, we have to convert from English words to mathematical terms. In terms of its width, we are asked to convey the length of the court. An algebraic equation showing the length "L" written in terms of the width "W" must be created.

The translation is here:

L = 44 + W W = 44

"The length" is "more than" the width of 44 feet.

So, the algebraic expression L = W+44 is the answer to our word problem.

Now, just for laughs and giggles, after defining keywords, let's look at a few more basic examples of translating from English to mathematical equations.

Example # 1 Write as an algebraic expression the number of y and 16.

"You can write this as" y + 16.

Example # 2 Write as an algebraic expression the distinction between 2x and y.

"You should write this as" 2x-y.

Example # 3 Write the ratio of 6 as an algebraic expression more than two times y to x.

"You should write this as" $(2y + 6) / x$

Developing the capacity to recognize keywords and translate phrases and sentences into mathematical equations is the secret to solving word problems. This ability can only be sharpened after doing lots and lots of these topics, as with anything else in mathematics. "After that, you will be secure in your ability to solve it if you come across a word problem instead of staring at it and thinking to yourself .... "Word Problems ...

Odd Or Even Numbers

As you continue to help your child get more comfortable counting, understanding what numbers mean, the concepts of "combining" numbers, and "take away," we will revisit a topic we discussed before: number comparisons. We're going to start looking at patterns of numbers as well.

Number comparisons:

Back in the beginning, we dealt just a little with number comparisons. Now that your child has a better grasp of numbers and can count farther, get out another number line (masking tape?) but make this one longer than before since your child knows more numbers. First, a little "check for understanding" to see if your child still remembers comparing one number to another. Is six more or less than 8? If your child remembers, yea!! If not, review a few problems on the number line. Remember, greater is "to the right of " and less than is "to the left of." Eight is greater than six because 8 is to the right of 6; or, six is less than eight because 6 is to the left of 8."

Now that your child is more verbal, you can begin to ask questions like: "Why is 15 less than 18?" or "How do you know that 30 is greater than 29?" If your child can speak in complete sentences with good understanding, then work on complete sentence answers. Do several of these until your child has the concept and practice these periodically.

Beginning number patterns:

Two very important number patterns that you can introduce your child to now are the even numbers and the odd numbers. You would be shocked at how many freshmen cannot explain what makes a number even or odd. They can tell you that even numbers end in a 0, 2, 4, 6, or 8, but they can't tell you that even numbers can be divided by 2 with no remainder. Likewise, when asked to algebraically "represent" a number that is even or odd, they use x. But x can represent any number. 2x, on the other hand, is always even and 2x + 1 is always odd.

To make this concept concrete, we need the whiteboards, the number line, and the objects you used for addition. Write the following on the whiteboards. Even numbers: 0, 2, 4, 6, 8, 10, 12, 14, up to about 20.

Under the evens, write the odd numbers. Odd numbers: 1, 3, 5, 7, 9, 11, 13, 15, 17, 19. Now have your child study (really look at and say them out loud) the even numbers while looking at and pointing to them on the number line. Then repeat for the odd numbers. Ask your child what he/she notices about these numbers.

Is there anything special about them?

Your child might notice that evens and odds alternate on the number line or that you have to "skip over" the evens to get the odds and vice versa. Your child might not have any idea what you are talking about or what you want. If this seems to be the case, point these things out. "Look. (sound semi-excited) as you go up the number line saying even or odd, they alternate. Even. Odd. Even. Odd." Have your child say that out loud. Also, point out the "skipping over" pattern. Think about anything else your child says. Is it true? Sometimes they see things we don't see because they aren't as focused. If they say something true, "Wow, I hadn't noticed that!" If what they say is not true, "That was a good try, but here's what I see."

We need those additional items because we want to get across the concept that evens can always be separated into groups of 2. Ask your child to count out a pile of 18 coins. Then ask to have them separated into separate groups of 2. Were there any "left over?" How many groups of 2? Write on the whiteboard: 18 is 9 groups of 2. Ask your child to pick another even number and repeat this process. Put into separate groups of 2. Any leftover? How many groups of 2? Write in on the whiteboard.

(This "groups of" talking is introducing the concept of multiplication. Just don't say that.)

Now, ask your child if he/she thinks this--separate groups of 2 with nothing left over--is true for every even number? Test as many events as it takes for your child to say YES. Even numbers always separate into groups of 2 with nothing remaining. They separate evenly. (This is beginning the concept of division and remainders, but don't say that either.)

What about the odd numbers? Pick one--7. Separate into groups of 2. Oh! Still, 1 leftover. Write on the whiteboard: 7 is 3 groups of 2 + 1. "That's odd." "Let's try another. 15 is 7 groups of 2 + 1. There it is again. Is this--separating into groups of two but having one leftover--always true for odd numbers? Practice until your child is sure.

Erase the whiteboard for a summary. Write out the even numbers up to 20 and these two observations: (1) even numbers always separate into even groups of 2, and (2) even numbers always END in a 0, 2, 4, 6, or 8. Now ask a few questions like "Is 36 even?" Yes, because it ends in a 6. Is 17 even? No, it doesn't end in 0, 2, 4, 6, or 8.

Now, add to the whiteboard the odd numbers to 19 and these two observations: (1) odd numbers cannot be separated into even groups of 2 because there is always one leftover, and (2) odd numbers END in 1, 3, 5, 7, or 9. Is 24 an odd number?

 NO, it doesn't end in 1, 3, 5, 7, or 9. Is 21 an odd number? Yes, it ends in a 1.

Do you need to do all of this in one day? No. Make the number comparisons one day. Evens another day. Or do a tiny bit each day. Just keep things easy, and your child successful.

Will your child remember this tomorrow? Some will, some won't. Don't expect it and you might get surprised. The concepts of even and odd are important, so keep practicing. When your child is answering questions correctly without using the coins, you can start looking for even and odd numbers where you go. But don't do that until your child has a very concrete understanding of even and odd.

Even Number

An even number is an integer of the form in which an integer is placed.

Thus, the even numbers are ... 0, 2, 4, 6, 8, 10, ... (OEIS A005843). Since even numbers are divisible by two in their entirety, unity holds for even. A single even number is called an even number for which it also holds, while a doubly even number is called an even number. An integer, which is not even is considered an odd number.

Its parity is called the oddness of a number, so an odd number has parity 1, while an even number has parity 0.

The even numbers' generating function is

As can be seen, by writing, the product of an even number and an odd number is always even.

Which is divisible by two and is even, thus.

How to Solve Simple Math Word Problems

Frank Howard Clark said, "I think finding some humor in it is the next best way to solve a problem." Mr. Clark would like to express in this that it is by having fun with the problems to solve simple math word problems. And in having fun with them, there are two simple steps to solving problems with math words:

Stage 1: Transform the words into an equation or numeric expression

It is possible to transform math word problems into a sequence of expressions or equations containing a mathematical expression combination. You have to take these steps to be able to translate these word problems:

1. Read very well and, in its entirety, the problem. Get the full viewpoint of the problem. Reading it in full will give you an idea of what the real problem is.

2. List all considerations mentioned. Make a list of all the variables given, including, if available, measurement units. If you need to do some conversion, such as miles to kilometers, pounds to kilograms, etc., you will be shown if all this information is available.

3. Defines what needs to be responded to. Make sure you know what you're searching for or what you need to solve the problem.

4. Have your solution prepared. To find the solution to the problem, include the procedures or

steps you will take. It will help you monitor all the variables and expressions you are using by showing the step-by-step process.

5. Be mindful of the keywords. In translating and solving the problem, when translating words into algebraic equations, you should be aware of the fundamental keywords, such as:

Addition: added to, increased by, more than, a sum of, a total of, combined with

Subtraction: decreased by, subtracted from, less than, a difference of, reduced by, fewer than

Multiplication: multiplied by times, a product of

Division: divided by, the quotient of, the remainder of, percentage, the ratio of, per

Certain key words suggest specific mathematical operations that should be done to the given factors or variables.

6. Plot the expression or equation. Plot the expressions or equations properly following the order of operations.

Step 2: Solve the math equation.

Follow the order of operations by stage to overcome a mathematical equation:

Calculate all those within parentheses or the innermost expressions first

Calculate those with exponents elevated to strength or root of

Multiplied or divided from left to right

Adding or subtracting from left to right

Writing down each level's answers before going to the next level would be simple to solve the equation. Here is an example:

$X = ((2 * 3) + (32) + (20/4) - (2 * 6)) * 2 + (3 * 8) - (4 * 5)$

$X = ((6) + (9) + (5) - (12)) * 2 + (24) - (20)$

$X = (6 + 9 + 5 - 12) * 2 + (24 - 20)$

$X = (8) * 2 + (4)$

$X = 8 * 2 + 4$

$X = 16 + 4$

$X = 20$

You will follow the order operations that were followed in solving the equation from the sample equation above.

There is no need for you to stop and stare and pray for divine intervention to solve a math word issue! You can confidently assume that math is simple and that math word issues are not difficult to solve.

How to Work on Math Problems

There is a five-step process for solving problems in mathematics that works at nearly every level. The steps are: diagram, formula, substitute, simplify, and solve. These steps also work for physics, chemistry, and engineering. Having a plan and implementation steps will greatly improve your math solving ability and make you more productive.

Diagram

First, draw a diagram to have a visual reference; the expression 'a picture is worth a thousand words' is especially true in mathematics. You may need to skip this step in algebra, but it is the most important step in geometry, trigonometry, or engineering. Always have the bigger picture in mind as it will help you to piece together the puzzle.

Formula

Next, choose an applicable formula, or you may need to convert a statement from English into mathematical notation. The formula gives you something to work with.

(Substitute)

Substitute known values, constants, or expressions into the formula from step two. Always substitute inside the parenthesis.

Simplify

As the old saying goes, please keep it simple. Simplify the resulting equation or expression following the order of operations. PEMDAS - parenthesis; exponents or roots, multiplication or division, and finally addition or subtraction.

Solve

Solve the equation. To solve algebraically: do the opposite, to both sides, in reverse order.

Even if one or two of the steps do not apply to your problem, they will usually need to be completed in the order that they appear above. The key argument is that to be more effective and have a better view of the bigger picture, and you need an outline.

CHAPTER 4: MATH WORD PROBLEMS, CATEGORIZED BY SKILL

Addition

1. Adding to 10: Ariel was playing basketball. One of her shots went in the hoop. 2 of her shots did not go in the hoop. How many shots were there in total?

2. Adding to 20: Adrianna has ten bits of gum for her mates to share with her. All her friends didn't have enough gum, so she went to the store to get three more pieces of gum. How many gum bits does Adrianna have right now?

3. Adding to 100: There are 10 bits of gum for Adrianna to share with her peers. All of her friends didn't have enough gum, so she went to the store and got 70 pieces of strawberry gum and 10 pieces of bubble gum. How many gum bits does Adrianna have right now?

4. Adding Just over 100: There are 175 regular chairs and 20 baby chairs at the restaurant. How many seats in total does the restaurant have?

5. Adding to 1,000: How many cookies have you been selling when you sold 320 cookies of chocolate and 270 cookies of vanilla?

6. Adding to and above 10,000: Normally, the hobby store sells 10,576 trading cards per month. The hobby shop sold 15,498 more trading cards than average in June. Overall, how many playing cards did the June hobby shop sell?

7. 3 Numbers added: Billy had two books at home. He went to the library and checked out two other books.

 He purchased one book then. How many books is Billy now carrying?

8. Adding three and more than 100 numbers: Ashley bought a large bag of sweets. There were 102 blue candies, 100 red candies, and 94 green candies in that bag. How many candies in total were there?

Subtraction

9. Subtracting to 10: The pizza shop had three pizzas in all. A customer ordered 1 pizza. How many remaining pizzas?

10. Subtraction to 20: Your friend says she's got 11 stickers. Just when you helped her clean her desk, did she have a total of 10 stickers. How many missing stickers are there?

11. Subtracting to 100: To share with her relatives, Adrianna has 100 bits of gum. She shared 10 bits of strawberry gum when she went to the park. Adrianna shared a further 10 bits of bubble gum as she left the park. How many gum bits does Adrianna have right now? [caption id= "attachment 2293" align= "aligncenter" width= "640"]

 Schools that use Prodigy consistently outperform those who do not use formal evaluations[/caption] by making math engaging.

12. Subtracting Slightly over 100: Your squad scored a total of 123 points. In the first half, sixty-seven points were scored. In the second half, how many were scored?

13. Subtracting 1,000: Nathan has a huge farm of ants. He agreed that some of his ants would be sold. With 965 ants, he began. He got 213 sold. How many ants has he got now?

14. Subtracting to and over 10,000:

the hobby shop usually sells 10,576 trading cards every month. The hobby shop sold a total of 20,777 trading cards in July. Compared with a regular month, how many more trading cards did the hobby shop sell in July?

15. 3 Numbers Subtracting: Charlene had a pack of 35 pencil crayons. She gave 6 to Theresa, her neighbor. She gave 3 to Mandy, her neighbor. How many pencil crayons were left by Charlene?

16. 3 Numbers to and beyond 100: Ashley bought a big bag of candy to share with her friends. There were 296 candies in all. She supplied Marissa with 105 candies. She also provided Kayla with 86 candies. How many other candies are left?

Multiplication

17. 1-Digit Integers Multiplying: Adrianna wants to break a pan of brownies into bits. In the pan, she cuts six even columns and three even rows. How many brownies has she got?

Eighteen. 2-Digit Integers Multiplying: A movie theatre has 25 rows of seats with 20 seats in each row. How many seats in total are there?

19. Multiplying Integers Ending with 0: There are four different sweatshirts for a clothing company.

The business makes 60,000 of each kind of sweatshirt each year. How many sweatshirts per year does the business make?

20. Multiplying three integers: Bricks are stacked in 2 rows by a bricklayer, with 10 bricks in each row. There is a stack of 6 bricks on top of each row. How many bricks in total are there?

21. Multiplying 4 Integers: By distributing newspapers, Cayley receives $5 an hour. Three days a week, for 4 hours at a time, she delivers newspapers. After delivering newspapers for eight weeks, how much money will Cayley earn?

Division

22. Dividing 1-Digit Integers: How many pieces of candy are in each bag if you have four pieces of candy divided equally into two bags?

23. Dividing 2-Digit Integers: How many rides would you go on if you have 80 tickets for the fair, and each ride costs five tickets?

24. Dividing Numbers Ending with 0: For buying new computer equipment, the school has $20,000.

How many items will the school purchase in total if each piece of equipment costs $50?

25 Dividing 3 Integers: For a total of $12, Melissa purchases two packs of tennis balls. There are 6 tennis balls altogether. How much does it cost for one bag of tennis balls?

? How much does it cost for 1 tennis ball?

26. Interpreting Remainders: A shipment of 86 veal cutlets is issued to an Italian restaurant. If it takes three cutlets to make a salad, after making as many dishes as possible, how many cutlets would the restaurant have leftover?

Operations Mixed

27. Mixing Addition and Subtraction: The library has 235 books. There were 123 books taken out on Monday. Fifty-six books were carried back on Tuesday. Now, how many books are there?

Twenty-eight. Mixing Multiplication and Division: A group of 10 people ordering pizza is open. How many pizzas do they order if they get 2 slices, and every pizza has 4 slices?

29. Multiplication, Addition, and Subtraction mixing: Lana has 2 bags in each bag with 2 marbles.

With 3 marbles in each bag, Markus has 2 bags. How many are more marbles there for Markus?

30. Division, addition, and subtraction mixing: Lana has 3 bags with the same number of marbles, 12 marbles in all. Markus has three bags, totaling 18 marbles, with the same number of marbles in them. How many more marbles does each bag have for Markus?

Ordering and Meaning Number

31. Counting to Preview Multiplication: In your classroom, there are 2 chalkboards. How many pieces do you need in total

if every chalkboard requires 2 pieces of chalk?

32. Counting to Preview Division: Your classroom has three chalkboards. There are two chalk bits on each chalkboard. This means that there are a total of

Six chalk bits. How many in total will there be if you removed one piece of chalk from each chalkboard?

33. Composing Numbers: What are 6 and 10 tens of numbers?

34. Guessing Numbers: In the Tens, I've got a 7. In one spot, I have an even number. I'm less than 74 now. What's my number?

35. Finding the Order: Mitchell scored more points than William in the hockey game, but fewer points than Auston. Who has the most points scored? Who had the fewest points scored?

Fractions

36. Finding Fractions of a Group: For Halloween, Julia went to 10 houses on her street. Five of the houses provided her with a chocolate bar. What fraction of Julia's street houses offer her a bar of chocolate?

37. Unit Fractions Finding: Heather is drawing a picture of Lisa, her best friend. She divides the portrait into 6 equal sections to make it simpler. Every part of the portrait represents what fraction?

38. Adding fractions with identical denominators: Each day, Noah walks 1/3 of a kilometer to school. He also runs 1/3 of a kilometer after school to get home. How many kilometers in total does he walk?

39. With Like Denominators, Subtracting Fractions: Whitney counted the number of juice boxes she had for school lunches last week. She'd got 3/5 of a

scenario. It is down to 1/5 of a case this week. How much did Whitney drink in this case?

40. Adding Whole Numbers and Percentages with Like Denominators: 6 1/4 scoops of chocolate ice cream, 5 3/4 scoops of vanilla, 2 3/4 scoops of strawberry served at lunchtime in an ice cream parlor. How many ice cream scoops in total did the parlor serve?

41. Whole Numbers and Fractions Subtracted with Like Denominators: Jaime had 5 1/3 bottles of cola for her friends to drink for a picnic. She drank 1/3 of a glass. Her mates were drinking 31/3. How many bottles of cola are left for Jaime?

4. Adding Unlike Denominators to Fractions: Kevin completed 1/2 of a school task. He completed 5/6 of another assignment while he was home that evening. How many tasks did Kevin finish?

43. Subtracting Fractions with Unlike Denominators: Patty used 7/8 of a ham box to prepare school lunches for her kids. She used 1/2 of a turkey kit as well. How much more ham has Patty made use of than turkey?

44. Multiplying Fractions: The students raced for 1/4 of a2 kilometer during gym class on Wednesday.

They ran half as many kilometers on Thursday as they did on Wednesday. On Thursday, how many kilometers did the students run? Write a fraction of your response.

45. Dividing Fractions: 1⁄5 of a bottle of color dye is used by a fabric maker to produce one pair of pants. 4⁄5 of a bottle was used by the manufacturer yesterday. How many pairs did the maker make of the pants?

46. Whole Numbers Multiplying Fractions: This week, Mark drank 5⁄6 of a carton of milk. Frank drank more milk than Mark 7 times. How many milk cartons did Frank drink? As a fraction, or as a whole or mixed number, write your answer.

Decimals

47. Adding Decimals: In your cup, you've got 2.6 grams of yogurt and added another spoonful of 1.3 grams. How much yogurt have you got in total?

48. Subtracting Decimals: To make a cake, Gemma had 25.75 grams of frosting. She determined that only 15.5 grams of frosting would be used. How much frosting is left for Gemma?

49. Multiplying whole numbers with decimals: every day, Marshall walks a total of 0.9 kilometers to and from school. After four days, how many kilometers will he walk?

50. Dividing Decimals by Whole Numbers: Mrs. Robinson obtained 2.5 kilograms of spaghetti to make the Leaning Tower of Pisa from spaghetti. Her students were able to make a total of ten learning towers. How many kilograms of spaghetti does 1 leaning tower take to make it?

51. Mixing Addition and Subtraction of Decimals: Rocco has 1.5 liters of orange soda and 2.25 liters of grape soda in its refrigerator. Antonio has an orange soda of 1.15 liters and a grape soda of 0.62 liters. How much is more soda there for Rocco than Angelo?

52. Mixing Multiplication and Decimals Division: Laura performs martial arts for 1.5 hours four days a week. What is her average practice time per day every week, considering a week is seven days?

Sequencing and Contrasting

53. 1-Digit Integer Comparison: You've got three apples, and your mate has 5 apples. Who's got more?

54. 2-Digit Integer Comparison: You've got 50 candies, and your mate has 75. Who's got more?

55. Comparing Various Variables: The playground has 5 basketballs. There are seven soccer balls in the playground. Are there any more soccer balls or basketballs?

56. 1-Digit Integers Sequencing: Erik has 0 stickers. He gets one additional sticker per day. How many days are there before he gets three stickers?

57. Skip-Counting by Strange Numbers: Natalie started at 5. She skipped-counted with fives. Could she have said the 20th number?

58. Natasha started at 0.0. Skip-Counting by Even Numbers: She skip-counted eight of them. Had she been able to say the number 36?

59. Sequencing 2-Digit Numbers: Jeremy adds the same number of cards to his set of baseball cards every month. He had 36 of them in January. Forty-eight in February. 60 from March. In April, how many baseball cards will Jeremy have?

Money and Time

60. Adding Money: Thomas and Matthew are saving money together to purchase a video game.

Thomas saved 30 million. Matthew saved 35 dollars. In sum, how much money did they save together?

61. Subtracting Money: Thomas saved $80 for himself. He uses the money he receives to purchase a video game. It costs $67 for a video game. How much cash has he left?

62. Multiplying Money: For delivering the document, Tim gets $5. After delivering the paper three times, how much money would he have?

63. Dividing Money: To buy three hockey sticks, Robert spent $184.59. How much did 1 cost, if. Was hockey stick the same price?

64. You went to the supermarket and purchased gum for $1.25 and a sucker for $0.50. Adding Money with Decimals: How high has your average been?

65. Subtracting Decimals from Money: You went to the store with $5.50. You purchased gum for $1.25, a $1.15 chocolate bar, and a $0.50 sucker. How much money have you got left?

66. Converting Hours into Minutes: For 1 hour, Jeremy helped his mom. How many minutes had he been helping her?

67. akob wants to invite 20 friends to his birthday, which will cost his parents $250.

Applying Proportional Relationships to Money: How much money would it cost his parents to invite 15 friends instead? Assume the relationship is proportional directly.

68. Applying Money Percentages: Retta deposited $100.00 in a bank account that pays 20% interest annually. In 1 year, how much interest will it accumulate? And if she does not make any withdrawals after one year, how much money is in the account?

69. Time to Add: If you wake up at 7:00 a.m. And it takes 1 hour and 30 minutes for you to get up and walk to school, and what time are you going to get to school?

70. Time Subtraction: If a train leaves at 2:00 p.m. And arriving at 4 p.m., how long have the passengers been on the train?

71. 71. Finding Start and End Times: At twenty to seven in the evening, Rebecca left her dad's store to

go home. She was home forty minutes later. When she got home, what was the time?

Physical Measurement

72. Comparing Measurements: The ruler of Cassandra is 22 centimeters long. The April Ruler is thirty inches tall. How many inches long is the April Ruler?

73. Measurements Contextualizing: Imagine a school bus. Which measurement unit will better represent the bus length? Centimeters, or kilometers, or meters?

74. Adding Measurements: Micha's father wants to save gas money, so he's been monitoring how much he uses. Micha's dad used 100 liters of gas last year. Her dad used 90 liters of gas this year. How much gas did he use for two years in total?

75. Measurements Subtracting: Micha's father wants to save money on petrol, so he's been monitoring how much he uses. Micha's dad has used 200 liters of gas over the last two years. He used 100 liters of gas this year. Last year, how much gas did he use?

76. Volume and Mass Multiplying: Kiera likes to make sure she has healthy bones, so she drinks 2

liters of milk every week. How many liters of milk will Kiera drink after three weeks?

77. Amount and Mass Dividing: Lillian does some gardening, so she purchased 1 kilogram of soil. She needs to divide the soil between her two plants equally. How much is each plant going to receive?

78. Converting Mass: Inger goes to the grocery store to buy three 500-gram-weight squashes. How many kilograms of squash has Inger purchased?

79. Converting Volume: Shad sells 20 cups of lemonade and has a lemonade stand. It was 500 milliliters per cup. How many liters in total did Shad sell?

80. Duration Converting: Stacy and Milda compare their heights. Stacy is 1.5 meters high. Milda is taller than Stacy by 10 centimeters. What is the height of Milda, in centimeters?

81. Distance and Direction comprehension: A bus leaves the school to take students on a field trip. The bus travels 10 kilometers south, 10 kilometers west, 5 kilometers south, and 15 kilometers north. Which way does the bus have to travel to return to school? How many kilometers in that direction does it travel?

Percentages and Ratios

82. Finding a Missing Number: The ratio of trophies for Jenny to Meredith's trophies is 7:4. There are 28 trophies for Jenny. How many have Meredith got?

83. Finding Missing Numbers: The ratio of trophies for Jenny to trophies for Meredith is 7:4. 12 is the difference between the numbers. What numbers are there?

84. Comparing Ratios: There are ten saxophone players and 20 trumpet players in the school's junior band. There are 18 saxophone players in the senior school band and 29 trumpet players. What band has a higher trumpet to the sax player ratio?

85. Determining Percentages: In her school, Mary surveyed students to find out about their favorite sports. Out of 1,200 pupils, 455 said their favorite sport was hockey. What percentage of students said their favorite sport was hockey?

86. Determining Percent of Change: Oakville's population was 67,624 individuals a decade ago. It is 190 percent bigger now. What is the actual population of Oakville?

87. Determining Percents of Numbers: 60 percent of 120 skates are for boys at the ice skate rental stand. How many are there if the majority of the skates are for girls?

88. Calculating Averages: William volunteered as a helper for swimming lessons for four weeks. He volunteered for 8 hours during the first week. He volunteered in the second week for 12 hours and in the third week for another 12 hours. He volunteered for 9 hours in the fourth week. How many hours, on average, did he volunteer a week?

Probability and Data Relationships

89. Understanding the Probability Principle: John wants to know the favorite TV show in his class, so he surveys all the boys. Would it be reflective or biased in the sample?

90. Understanding Tangible Probability: 1, 2, 3, 4, 5, and 6 are labeled as the faces on a fair number die. Twelve times, you roll the die. How many times do you have to expect a 1 to roll?

91. Complementary Events Exploring: The numbers 1 to 50 are in a hat. If the probability of drawing an even number is 25/50, what probability

will an even number not be drawn? Express this chance as a fraction.

92. Exploring Experimental Probability: Recently, a pizza shop sold 15 pizzas. 5 of those were pepperoni pizzas. Answering with a fraction, what is the experimental

 possibility of pepperoni being the next pizza?

93. Data Relationships Introduction: Maurita and Felice take four tests each. Here are the four experiments carried out by Maurita: 4, 4, 4, 4. For 3 of Felice's four studies, here are the results: 3, 3, 3. If Maurita's average is 1 point higher than Felice's for the four tests, Felice's 4th test score?

94. Introducing Proportional Relationships: For $7.00, Store A sells 7 pounds of bananas. For $6.00, Store B sells 3 pounds of bananas. What store will have a better deal?

95. Proportional Relationship Writing Equations: Lionel loves soccer but has difficulty motivating himself to practice. So, by video games, he stimulates himself. The number of drills that Lionel conducts, in x, and how many hours he plays video games, in y, have a proportional relationship. He plays video games for 30 minutes when Lionel

completes ten drills. For the relationship between x and y, write the equation.

Geometry

96. Introducing Perimeter: four chairs in a row are in the theatre. Five rows are here. What is the perimeter, using rows as your unit of measurement?

97. Introducing Area: Four chairs in a row are in the theater. Five rows are here. How many seats in total are there?

98. Amount introduction: Aaron needs to know how much candy his container can hold. The container is 20 inches tall, 10 inches long, and 10 inches wide. What is the volume of the container?

99. 2D Shapes understanding: Kevin draws a shape with four equal sides. What form was he drawing?

100. Finding the Perimeter of 2D Shapes: On a square paper piece, Mitchell wrote his homework questions. The paper is 8 centimeters on each side. What perimeter is there?

101. Area of 2D Shapes Determination: A single trading card is 9 centimeters long by 6 centimeters high. What is its territory?

102. 3D Shapes understanding: Martha draws a shape with 6 square 3D Shapes understanding: Martha draws a shape that has 6 square What form was she drawing?

103. Determination of 3D shapes' surface area: What is the surface area of a cube 2 cm wide, 2 cm high, and 2 cm long?

104. 3D Shapes Volume Determination: Aaron's container of candy is 20 centimeters tall, 10 centimeters long, and 10 centimeters wide. The container for Bruce is 25 centimeters high, 9 centimeters long, and 9 centimeters wide. Find each container's volume. Whose container should hold more sweets, based on volume?

105. Right-Angled Triangles Identification: A triangle has the following side lengths: 3 cm, 4 cm, and 5 cm. Right-angled is this triangle?

106. Equilateral Triangles Identification: A triangle has the following side lengths: 4 cm, 4 cm, and 4 cm. What type of triangle is that?

107. Identifying Isosceles Triangles: The following side lengths of a triangle are 4 cm, 5 cm, and 5 cm. What type of triangle is that?

108. Scalene Triangles Identification: A triangle has the following side lengths: 4 cm, 5 cm, and 6 cm. What type of triangle is that?

109. Finding the Perimeter of Triangles: In the form of an equilateral triangle, Luigi constructed a tent. The perimeter measures 21 meters. What is the length of each side of the tent?

110. Triangle Area Determination: What is the area of a triangle with a base of two units and a height of three units?

111. Applying the Pythagorean Theorem: One non-hypotenuse side length of 3 inches is a right triangle, and the hypotenuse measures 5 inches. What is the other non-hypotenuse side length?

112. Finding the Diameter of a Circle: Jasmin purchased a new round backpack. Its area is 370 centimeters square. What is the diameter of the round backpack?

113. Finding a Circle's Area: The circular shield of Captain America has a diameter of 76.2 centimeters. What is his shield area?

114. Finding the Radius of a Circle: Skylar lives on a farm where his dad keeps a circular corn maze.

The diameter of the corn maze is 2 kilometers. What is the radius of the maze?

Variables

115. Identifying Independent and Dependent Variables: Victoria, for her class, is baking muffins. The amount of muffins she makes depends on how many classmates she's got. M is the number of muffins in this equation, and c is the number of classmates. Which is an independent variable, and which is a dependent variable?

116. Writing Addition Variable Expressions: Last football season, Trish scored goals for g. Alexa scored four more than Trish's goals. Write an expression that reveals how Alexa scored several goals.

117. Writing Variable Subtraction Expressions: Elizabeth eats a good, nutritious breakfast every week. Madison skips breakfast often. Madison eats three fewer breakfasts per week in total than Elizabeth. Write an expression that illustrates how Madison eats breakfast several days a week.

118. Writing Multiplication Vector Expressions: Last hockey season, Jack scored g objectives. Patrik scored twice as many goals as Jack. Write

expressions that indicate how many goals were scored by Patrik.

119. Writing Variable Division Expressions: Amanda has chocolate bars with c. She wants to divide the chocolate bars equally between her three mates. Write an expression that indicates how many chocolate bars will be offered to 1 of her mates.

120. Solving Two-Variable Equations: This equation shows how Lucas's sum depends on how many hours he works from his after-school job: $e = 12h$. How many hours he works is expressed by the variable h. How much money he receives is expressed by the variable e. After working for 6 hours, how much money would Lucas earn?

How to make your own Math Word issues simple

Armed with 120 examples to inspire ideas, you will involve your students and maintain continuity with lessons by making math word problems. Do - Do:

Link to Student Interests: You will catch attention by framing your word problems with student interests. For instance, if most of your class loves American football, the throwing distance of a popular quarterback might involve a measurement issue.

Make Topical Questions: Writing a word problem representing current events or problems will motivate students by providing them with a simple, concrete way to apply their information.

Include Student Names: Naming the characters of a question after your students is a simple way to make the subject matter relatable and help them work through the topic.

Be explicit: The query is distilled by repeating keywords, helping students concentrate on the core issue.

Don't:

Comprehension of Test Reading: Flowery word choice and long sentences can mask the main elements of a question. They are using concise phrasing and grade-level vocabulary instead.

Focus on common interests: Framing too many questions with related interests can alienate or disengage some students, such as football and basketball.

Feature Red Herrings: Another problem-solving factor is added by adding needless details, overwhelming many elementary students.

Word problems that students can connect to and contextualize can capture attention rather than generic and abstract ones, a key to differentiated teaching.

Final Thoughts on Math Word Problems

Using the issues as templates, you'll possibly get the most out of this resource, slightly changing them by adding the above tips. In doing so, they can be more important to the students — and interact with them.

CHAPTER 5: ELEMENTARY ALGEBRA ARE DISTANCE PROBLEMS

Word concerns are a perfect way for students to find applications in the real world for the knowledge they are studying in the classroom while improving analytical abilities for thinking. Analyze how you can solve it yourself and settle about the correct way for your students to write a word problem.

Since word issues sometimes have a plot of some kind, they are often referred to as story issues and can differ in the amount of technical language used.

Distance problems, age problems, job problems, percentage problems, combination problems, and number problems are the most common word problems in elementary algebra.

A typical age problem:

Ann is three times as old as Bob, her baby brother. She will only be half as old in five years. Now, how old are they?

One first transforms the terms into logical variables, operations, and equations to overcome this by algebra:

• Write the age of Ann as variable A and the age of Bob as B.

Their ages are A+5 and B+5 five years from now.

Twice as old suggests that one generation equals two times the other, and three times in the same way.

The problem thus becomes:

For variables A, B, solve equations A = 3B and A+5 = 2(B+5).

The answer is A = 15, B = 5, or in the usual language: Ann is 15 years old, and Bob is five years old.

Structure, Structure

It is possible to analyze word problems such as the above on three levels:

A. The verbal formulation;

B. Mathematical relationships underlying them;

C. The Mathematical Symbolic Expression.

Such metrics as the number of terms in the problem or the mean sentence length may be used in linguistic properties. One system for analyzing logico-mathematical properties is to classify the numerical quantities in the problem into known quantities (values are given in the text), quantities wanted (values to be found), and auxiliary quantities (values to be found as intermediate stages of the problem)

Purpose and utilization

Mathematical modeling questions usually involve word problems, where data and knowledge about a certain system are presented, and a student is expected to create a model. For instance,

1. Jane had 5.00 dollars, then spent 2.00 dollars. How much has she got now?

2. The water is rising at a rate of 3 cm / s

in cylindrical barrel with a radius of 2 m. What is the rate of rising water volume?

These examples are intended to lead students to

independently create mathematical models, evaluate real-life scenarios, and encourage mathematical interest and understanding.

The importance of students of these issues varies.

From the sixteenth century onwards, the modern notation that allows mathematical ideas to be symbolically represented was developed in Europe. Before that, all mathematical issues and solutions were written out in words; the more complicated the problem, the more laborious and complex the verbal explanation was.

Examples of word problems dating back to Babylonian times can be identified. Most Old Babylonian concerns are couched in a language of measurement of daily items and events, aside from a few procedural texts for seeking things like square roots. Students had to find lengths of dug canals, stone weights, lengths of broken reeds, field areas, numbers of bricks used in building, etc.

Examples of word problems are also found in Ancient Egyptian mathematics. An issue is included in the Rhind Mathematical Papyrus that can be translated as:

There are seven houses; seven cats are in each house; each cat kills seven mice; each mouse has consumed seven grains of barley; seven heat would have been produced by each grain. What is the sum of all the items enumerated?

In more recent days, satire was the often confounding and subjective essence of word problems. This nonsensical problem, now known as the

 of the Captain, was written by Gustave Flaubert:

Since you are learning geometry and trigonometry now, I'm going to ask you a question. A ship sails across the ocean. With a cargo of wool, it left Boston. It'll gross 200 tons. For Le Havre, it is bound. The mainmast is broken, the cabin boy is on deck, 12 passengers are on board, the east-north-east wind is blowing, and the clock points to a quarter past three in the afternoon. It's in May. How old is a captain like this?

Word problems were also satirized in The Simpsons when a long word problem ("An express train leaves Santa Fe bound for Phoenix 60 miles per hour, 520 miles away. At the same time, instead of thinking that he's on the train, a local train traveling 30 miles an hour carrying 40 passengers leaves Phoenix bound for Santa Fe ...") trails off with a schoolboy character.

How to write equations based on word problems with algebra

I know you're always sitting in class, thinking, "Why am I forced to think about equations, algebra, and variables?"

Trust me, because there are circumstances in which you are going to use your Algebra skills to solve equations to solve a problem that is not relevant to the school. And ... And ... If you can't, you'll want to remember how you did it.

It might be a time when you want to find out how much you're going to get paid for a job, or maybe more importantly, whether you've been paid enough for a job you've done. It could even be when you try to find out whether you have been overcharged for a bill.

This is an important stuff-you would want to make sure that you get paid enough and not spend more than you have to when it comes time to spend YOUR money.

All right ... OK, well ... Let's put to work all of this newly acquired experience.

Let's look at an example of an issue with an algebra term.

Example 1: Word Problems in Algebra

Linda was selling school game tickets. She sold ten more tickets for adults than tickets for children, and she sold twice as many tickets for senior citizens as tickets for children.

1. Let x represent the number of tickets for children sold.

2. To indicate the number of adult tickets sold, write a phrase.

3. Write a term reflecting the number of tickets sold to seniors.

4. Tickets for adults cost $5, tickets for kids cost $2, and seniors' tickets cost $3. Linda made $ 700. To reflect the overall ticket sales, write an equation.

5. How many tickets for children were sold for the play? How many tickets for adults were sold? How were many tickets for seniors sold?

This problem is, as you can see, huge! With several phrases to compose, there are five questions to respond to.

The Solution

A few remarks concerning this issue

1. The variable was specified for you in this issue. Let x reflect the number of tickets sold by children to indicate what x stands for in this problem. You would have written it like this if this had not been done for you:

Let x = Number of tickets for children sold

2. I learned, for the first time, that ten more adult tickets had been sold. Since more means add, my phrase was x +10. I don't need an equivalent sign because the direction asked for an expression.

An equation is written with an equal sign, and there is no equal sign in an expression. We don't know the total number of tickets at this stage.

3. For the second phase, I realized that twice as many of my keywords meant two times as many. Therefore, my speech was 2x.

4. We know that we have to multiply each ticket's price by the number of tickets to find the total price. Please note that you have to place it in parentheses as x + 10 is the number of adult tickets! So, if you

multiply the price by $5, you're going to have to distribute 5.

5. I know the number of children's tickets until I solve for x, and I can take my expressions I wrote for # 1 and substituted 50 for x to find out how many adult and senior tickets have been sold.

The next example illustrates how a constant inside a word problem can be defined.

Example 2-Identifying a Constant

A mobile phone operator charges $12.95 and $0.25 a minute per call at a monthly rate. The M-minute bill is $21.20.

1. Write an equation that models this scenario.

2. On this bill, how many minutes is charged?

Solution

A mobile phone operator charges $12.95 and $0.25 a minute per call at a monthly rate. The M-minute bill is $21.20.

1. Write an equation that models this scenario.

Notes For Example 2 For example 2

The monthly rate is $12.95. I know that this is a constant because this is a fixed fee for each month. The rate does not change; it is not, therefore, linked to a variable.

$0.25 per minute per call needs a variable since, depending on the number of minutes, the overall sum would change. We use the expression 0.25 mm, thus,

To calculate the value of m, which is the number of minutes paid, you must solve the equation.

A word problem that needs an equation with variables on both sides is the last example.

Example 3-Both Sides Equations with Variables

You've got 60 bucks, and your sister has 120. You save 7 dollars a week, and your sister saves 5 dollars per week. How long is it going to be before you and your sister get the same amount of cash? Write and resolve an equation.

The Solution

Sample 3

$60 and $120 are constants since they both have to start with this amount of money. It does not adjust this number.

Tariffs are $7 a week and $5 per week. In this case, the main word 'per' means to multiply.

• In this issue, the keyword "same" means that I will set my two expressions equal to each other.

When the two expressions are set equal, we now have an equation on both sides with variables.

After solving the equation, you discover that x = 30, which means you and your sister will have the same amount of money after 30 weeks.

CHAPTER 6: SOLVE MULTIPLICATION WORD PROBLEMS

Are you good at problem-solving with words? Solving word problems is an important aspect of mathematics since you understand and practice what you have learned to do (add, subtract, multiply) about them.

We're going to look at issues with multiplication terms today: how we understand them, what to do to fix them ... all of which we're going to look at in this section.

1. Problems with word multiplication: Repetition

This is the first sort of problem with multiplication words that one knows how to do. For instance:

Anna has five cartons of eggs. There are 12 eggs in each carton. How many eggs in total does she have?

We'll find:

• A range of sets: there are five egg cartons for Anna.

• The number of items in each set: there are 12 eggs in each carton.

The question about the total amount of stuff is: how many eggs does she have in total?

To solve this word dilemma, we should think: if there are 12 eggs in each carton and Anna has five cartons, we will add $12 + 12 + 12 + 12 + 12 + 12$ to see how many eggs there are in total, or, what ends up being the same: we will multiply 5×12: Anna has 60 eggs in total.

2. Problems with word multiplication: One-step comparisons

One quantity is contrasted with another quantity that is greater or smaller in this form of word multiplication problem:

John put in $10 to buy his father's present, and Patricia put in 3 times more money than John. How much money was put in by Patricia?

We'll find:

A quantity representing one quantity: John has positioned $10.

A number representing the contrast between the second quantity and the first quantity: 3 times as much money as John was placed in by Patricia.

Question about the second amount: How much cash did Patricia put in?

We should think to solve this word problem: if Patricia put in 3 times as much money as Juan and put in triple the amount of money, we'll multiply 10 x 3. Patricia contributed $30.

3. Problems with multiplication words: One-step formulas

Formulas will appear to us in this type of multiplication word problem, e.g., a speed formula:

Justin is the driver of a bus. He told me that he would complete his route exactly 2 hours if he did not stop and still kept the same speed of 80 miles per hour. How many miles is the itinerary for him?

We'll find:

A pace: if a speed of 80 miles per hour has been maintained...

A time: He's going to arrive in 2 hours.

Distance question: How many miles is his route?

We should think to solve this word problem: if he maintains a speed of 80 miles per hour, that will mean he covers 80 miles per hour he travels. We also know that he drives for 2 hours at this speed. Therefore, we will have to add 80 x 2 to know the number of miles he travels in total: His path is 160 miles.

4. Problems with multiplication words: Mixture or Cartesian product

We'll find two or more sets of items or individuals in this form of the multiplication word problem. These sets should be combined, creating all possible pairs:

We went to eat at an Italian restaurant specializing in pasta today. Since there were nine types of pasta and 11 types of sauce on the menu, and you could mix any pasta with any sauce, it was hard for me to decide which dish to order. How many different combinations of pasta and sauce would you choose from?

We'll find:

• In the first set, the number of elements: 9 types of pasta.

• In the second collection, the number of elements: 11 kinds of sauce.

The question refers to the number of potential combinations between the sets: how many different combinations of pasta and sauce can you choose?

We ought to consider combining each type of pasta with 11 available sauce types to solve the word problem. Consequently, we can get 11 different dishes with only the first form of pasta. We could get 11 more plates by mixing the second form of pasta with each of the 11 sauces. We will get 11 distinct sauce variations for each of the nine types of pasta. We will then multiply 9 x 11 to find out the number of potential dishes: you may choose from 99 different pasta dishes.

These are the four primary templates for problems with word multiplication. We have a number of these types of word issues at Smartick and more. Register now if you wish to practice them!

Word Solving Problem

Word issues also confuse students simply because, in a ready-to-solve mathematical equation, the question does not present itself. Even the most difficult word problems can be solved if you understand the mathematical concepts discussed. Although the degree of difficulty which vary, solving word problems involves a planned approach that requires the problem to be defined, the relevant information collected, the equation generated, solved, and the work verified.

Identify the problem

The dilemma allows you to solve it by assessing the scenario. This could come as a challenge or a declaration. Either way, the word issue gives you all the data you need to solve it. You will decide the unit of measurement for the final response once you define the question. The question asks you to calculate the total number of socks between the two sisters in the following case. The measurement unit for this issue is the sock pairs.

"Suzy has eight pairs of socks in red and six pairs of socks in brown. Eight socks are owned by Suzy's brother Mark. If her little sister has nine pairs of purple socks and loses two pairs of Suzy's, how many pairs of socks are left for the sisters?"

Collect data

For any information, you don't know yet, create a table, list, graph, or chart that summarizes the information you know, and leave blanks. A different format might be required for each word issue, but a visual representation of the correct information makes it easier to deal with.

The query in the example asks how many socks the sisters own together so that Mark's data can be overlooked. Often, it doesn't matter the color of the socks. This removes much of the details and leaves you with only the total number of socks with which the sisters began and how much the little sister lost.

Creating an Equation

Translate all of the words of math into symbols in math. The terms and phrases "sum," "more than," "increased," and "in addition to" all mean to add, for instance, so write over these words in the "+" symbol. For the unknown variable, use a letter and construct an algebraic equation that represents the problem.

Compute the average number of sock pairs that Suzy owns, such as eight plus six. Take the total number of pairs her sister holds — nine. The total sock pairs held by both sisters are 8 + 6 + 9. Deduct the two missing pairs for a final equation of (8 + 6 + 9)-2 = n, where n is the number of pairs of socks left by the sisters.

Resolve the problem

Solve the problem using the equation by plugging in the values and solving for the unknown variable. To prevent any errors, double-check your calculations along the way. Using the order of operations, multiply, divide, and subtract in the proper order. Exponents and roots arrive first, then multiplication and division, and finally addition and subtraction.

In this case, you answer n = 21 pairs of socks after adding the numbers together and subtracting them.

Verify the Answer

Check if what you know makes sense in your answer. Estimate an answer, using common sense, and see if you come close to what you predicted. Check through the issue to see where you went wrong if the solution seems absurdly big or too short.

In that case, you know that you have a maximum of 23 socks by adding up all the numbers for the sisters. Since the problem mentions that the little sister lost two pairs, the final answer has to be less than 23. If you get a higher number, you've done the wrong thing. Apply this logic, regardless of the complexity, to every word query.

Let's say you have returned to math class. You will need to find the answers to the following questions: If you were given three pills by your doctor and ordered you to take one every half an hour, how long would it last for you? There are seven sisters in a certain family, and each sister has one brother. How many are those siblings? Or, if you divide 30 by half, how much do you have?

Not only is problem-solving one of the essential components of mathematics study, but it also permeates all facets of life, including the professional environment. Problem-solving encourages students to be out-of-the-box strategic thinkers, hone organizational skills, and create a rational reasoning method for reasoned decisions to be made. Problem-solvers will seek technological careers someday and become future researchers, inventors, designers, and engineers.

There is only one problem with problem-solving: When we look at the math segment of a standardized test such as the PARCC (Partnership for College and Career Readiness Assessment), particularly at the middle school level, where students' competency rates in grades 6-8 who met or exceeded math standards dropped below 35% in 2015-16, we can see that it is the word-problem portion that.

In part, because of their descriptive language, word problems appear to be complicated. Students often do not understand what they are being asked, especially when abstract concepts are included in the issue. Other difficulties occur when students lack the foundations of math and do not devise a strategy to solve or isolate an equation's steps.

The Plan for Problem-Solving

To spend hours planning and teaching bell-to-bell to see little to no progress can be discouraging. Similarly, students themselves get upset as their confidence erodes.

I have carried out standardized tests and witnessed students pressing the "skip" button or typing "idk" letters (short for I don't know) as soon as a multi-step word problem has been introduced to them. I have seen students give up on a two-hour segment after just 15 minutes. We'd discussed all the skills they needed to solve problems, but during testing, in silent despondency, I could do nothing but look on.

I also remember spending weeks in the classroom isolating a single mathematical subject and approaching it from every possible angle, only to present a question in the form of a word problem and have students reply with blank stares as the crickets chirped in the back row. It seems like the class could not piece together the measures required to think on their feet. I started to wonder: maybe the problem was not with the word problems themselves, but with difficulty teaching proper problem-solving strategies.

If we trace the mathematical roots back to earlier grade levels, we see that keywords often help younger students develop a problem-solving strategy. "We might use the word" more "as an example of what to add, for instance:" Three kittens curled up on a rug. On the blanket, four more kittens are crawling. How many kittens on the blanket are there now? Similarly, subtraction is implied by the term "less": "Marcy has six cats less than Nancy."

Similarly, subtraction is implied by the word 'fewer': Marcy has 12 cats. How many does Nancy have? The challenges occur when there is a lack of keywords to follow: "Some kittens were asleep on the blanket." Seven of them got up to have breakfast. Three are now on a blanket. How many on the blanket were there, to begin with? These types of fundamentals could be worth refreshing before diving into other tips and tricks. Each problem is unique, and there is no single, overarching algorithm for teachers to solve them all. So, how are we teaching students to read and translate a problem into math in English?

Practice Makes Perfect

My achievement was to introduce students to everyday representations of the process in more deliberate ways. Provide guided experience for students by learning various web problems from previously published math contests and standardized tests. Sort the issues into stages or topics of difficulty and begin with a one-step issue before moving to those with two or three steps. As students gradually become involved,

1. Read the problem

2. Understanding of what is being asked

3. Create a plan to fix the most challenging step

4. Execute the timetable

5. Verify how fair your solution is

Also, it would help if you never underestimated the machine as a strong instructional tool. By applying knowledge markers, defining variables, recognizing the unknowns in expressions, and describing their logic, students must learn to convert words into a sequence of steps toward a solution. One way to do this is to think about how problems can be attributed to their everyday lives.

The representation of mathematical concepts in diagrams may also promote a deeper understanding. To illustrate the visual relationship between the data and the unknown, use sketches, figures, or symbols. Build a conceptual map and outline in practice the necessary steps. Students should also have the ability to demonstrate their problem-solving strategies to others in the classroom as they learn the steps required.

These steps would provide students with the tools to find out that the pills that the doctor gave will last an hour (you take one pill instantly, the second pill in 30 minutes, and the last pill 30 minutes after that); that there are eight family siblings (the seven sisters share the same brother) and that 60 is what you get when you divide 30 by half.

Examples of word problems in mixed operations that can be resolved in three or more steps. In terms of the details provided and the data that needs to be found, we will explain how block diagrams can help you visualize word problems. Inside Singapore Math, block diagrams or block modeling are used. In Common Core, block diagrams are also called Tape Diagrams.

EXAMPLE:

Each crate had 42 mangoes. Twelve such mango crates have been shipped to a factory. There were four rotting mangoes, and they had to be thrown away. They packed the remaining mangoes into boxes of 10 mangoes each. How many boxes were there with the mangoes?

The Solution:

Phase 1: Find the number of mangoes delivered in total.

Word Problems in 3-step

$12 \times 42 = 504 = 504$

The number of mangoes delivered was 504 in all.

Phase 2: Find the number of mangoes that remain.

Problems with Words-mixed op

504 to 4 = 500

Phase 3: Find the number of mango boxes.

$= 500 \div 10 = 50$

50 boxes of mangoes were there.

EXAMPLE:

At the football game, there were 9500 spectators. Of these, six thousand and three hundred and seventy-five were citizens. There were four times as many children as women among the remaining spectators. How many babies were there?

The Solution:

Phase 1: Find the number of children and females.

9500-6375 = 31255 3125

Three thousand one hundred twenty-five(3125) women and children were there.

Phase 2: Find women's numbers.

$3125 \div 5 = 625 = 625$

625 women were there.

Phase 3: Find the children's number.

$625 \times 4 = 2500 =$

2500 kids were there.

CHAPTER 7: ADDITION AND SUBTRACTION WORD PROBLEMS

There are many types of word problems that can be faced by learners. There may be more variations for each of the various forms. For example, in word problems, the change in quantities, the starting number, or the final amount, or the amount of the change itself, may be unclear.

Students must learn to solve all these various kinds of problems, as this will demonstrate a complete understanding of the importance of addition and subtraction operations. Practice with many examples is needed, but this should be done with concrete materials after many hands-on activities.

The following provides examples of these various types of problems with addition and subtraction. Note: The table is based on the Common Situation of Addition and Subtraction

Various Addition and Subtraction Situations Styles

Append To There are three ducks on a pond. Four more ducks are arriving on the pond. How many ducks do they have now? 3 + 4 = 7

On a pond, three ducks are swimming. Some more ducks are coming in and landing next to them, and there are five ducks now. How many ducks were flying inside? 3 + = 7 = 7

On a pond, several ducks are floating. Next to them are four more ducks, and there are now seven ducks. How many of the ducks in the pond were there when the four landed? + 4 = 7 = 7

Seven candles were burning on a cake, Take From. Jake blew out 3 of them. How many were burning left? 7-3 = 4 = 4

Seven candles were burning on the cake. Some of them were blown out by Jake to leave four smoking. How many was he blowing out? 7- = 4 4

On the cake, there were some candles. Jake blew out four of them to leave three of them on fire. How many candles were already burning at the beginning? -4 = 3 = 3

Bring together/ Put together/

In a cup, you can find four red grapes and three white grapes. In the cup, how many grapes are there? For 4 + 3 =

In a bowl are seven grapes. Four of them are red, and the others are white. In the cup, how many white grapes are there? $4 + = 7 = 7$

$7-4 =$

Compare the three games for Sam and the seven games for Jack. How many more games are there for Jack? There are three ORSam games, and Jack has seven games. How many fewer games has Sam got? $3 + = 7 = 7$

$7-3 =$

Jack's got four games more than Sam does. Sam has three games. How many games has Jack got? There are four fewer games on ORSam than Jack. Sam has three games. How many games has Jack got? $3 + 4 = $ of

For $4 + 3 =$

Jack's got four games more than Sam does. Jack's got seven games. How many games has Sam got? There are four fewer games on ORSam than Jack. Jack's got seven games. How many games has Jack got? $7-4 =$

$+ 4 = 7 = 7$

Vocabulary and Vocabulary for Addition and Subtraction

It allows students to learn the vocabulary of addition and subtraction to solve each problem mentioned above. E.g., how many are required in total, total, combined, more than, difference, how many.

Both relied on and developed reading and language skills to solve word problems. Be conscious of your kid's reading level and use the ability to improve these abilities when you are working and discussing the issues. Help your children recognize the main words and phrases within an issue and understand them. Only a few examples are included in the table below.

Leave as much time for your children and offer support and motivation to help them interpret the issue. Work with them to understand the issue and decide the appropriate arithmetic operation to be converted into an equation of addition or subtraction.

When addressing issues, be concise and watch out for badly worded problems. 'Jack has seven console games, for instance, and Sam has four console games. How many altogether do they have?

"As" Jack has seven console games and Sam has four console games, it would be better worded. How many console games have they got in total? When more does not imply adding more,

When solving word problems, students search for verbal hints.

"More" typically implies addition (but not always), and "less" usually (but not always) suggests subtraction. Look out for issues where these words mean the opposite of what they normally do. The Green Team had 14 players, which was two more than the Red Team,

for instance. Students who can translate this as 14-2 = 12 are well on their way to knowing addition and subtraction, how many players were on the Red Team. Mistakes made with word issues

The following is based on Anne Newman1 's job.

It is possible to categorize errors that students create in word problems into one of five types:

• Reading of

Keywords or symbols are wrongly read.

Understanding-the inability to comprehend either the entire problem or particular parts of it

Transformation-incorrect recognition of the activities needed to solve the issue

Procedural or true-inappropriate estimate

Encoding-The solution is found but not specified correctly or entirely

The following helps to recognize what kind of mistake they are making, asking students to do or answer:

1. Read the question.

2. What was the thing that you were asked to do?

3. How were you planning on finding the answer?

4. Show me how you found the answer.

5. What was your answer?

Newman, 1977,' An study of the mistakes of sixth-grade pupils on written mathematical duties,' Victorian Institute of Educational Science Bulletin, vol 39, pp 31-43. Worksheets Addition / Subtraction Word Issue

There are plenty of word issues with the two worksheets below.

Please work with your kids through them and provide vocabulary support when needed.

DIVISION

Dividend, divider, quotient, and remainder are the terminology used in division. Repeated subtraction is division.

For example:

$24 \div 6$
How many times would you subtract six from 24 to reach 0?
24 - 6 = 18 one time
18 - 6 = 12 two times
12 - 6 = 6 three times
6 - 6 = 0 four times

There are two division facts for every multiplication truth.
For example:
1. $7 \times 5 = 35$
$35 \div 5 = 7$; $35 \div 7 = 5$

2. $6 \times 8 = 48$
$48 \div 8 = 6$; $48 \div 6 = 8$

3. 9 × 5 = 45 is the same as 45 ÷ 9 = 5

7 × 9 = 63 is the same as 63 ÷ 7 = 9

Below, each of the words used in the division is explained:
The dividend is called the amount, which is split.

The number that splits is called the splitter.

The quotient is called the number that is the product of the division.

It is called the remainder if there is any number left over.

CHAPTER 8: PUZZLE FOR KIDS

Are you having problems leaving your children at home alone? Is it still a concern when you have to go to work, and your child is at home alone? Oh, you can't think of what to give your child to do to keep him/herself occupied and not engage in other things that are not so effective. I'm intrigued by a nice time pass that can become your kid's best friend. You can find more than hundreds of puzzles for kids online or on the street today.

Puzzles for all age ranges are fantastic. Not only your baby. Part of your spare time may even be spent attempting to solve puzzles. They can be a good brain workout and can even make you relax at times. Puzzles are time-consuming at times, although others take approximately a minute. You might give your kid a puzzle when you go out for a long time that he/she needs about an hour to solve or otherwise give them little teasers they can do to brush up their brains.

If you look for puzzles for children online, you can find various puzzles in different categories. Puzzles involving words, numbers, crosswords, sketches, and popular jigsaw puzzles are available. Jigsaw puzzles can range from 5 to more than 1000 pieces. So, for all age ranges and all degrees of difficulty, they are fine. Word and number puzzles may help sharpen your kids' brain, and crosswords will help them develop vocabulary.

Most of you might be thinking of leaving a drawing or coloring book with your kids. Most children, however, draw on two pages, color two pictures, and then get bored. Puzzles hold them active. They don't quite remember how long it's been since they tried to fix it. And after one is done, they feel eager and excited to move on to the next to see how easily they can do it.

Another fascinating thing that makes it easier for you to keep kids' puzzle games is that they are readily accessible. If you are late for a party and don't have a puzzle book that you can get for your boy, log in to the internet, and you will be able to find a huge selection of printable puzzles for children. Just take a printout out and hand it over with a pencil to your boy. So, you have to buy a kid's puzzle for your little one!

It may be difficult for parents to find events that are not only enjoyable for their children but also educational. Including word searches to word unscrambles, there are plenty of puzzles that will force the children to use their brains.

Here are five excellent children's puzzles that you can either help them solve or spend hours learning on them with enthusiasm. Of course, it's crucial to make sure that the games you pick are suitable for the kid's age, whether they are classic board games or jigsaw puzzles.

Word Searches-The word search game has been popular for a long time, and as there are phases of progress, there are as many different levels that you can play with your boy. The great thing about looking for words is that the kid doesn't need to know what a word means in the puzzle to find it. And this leads to the second concept of puzzles that you can play with your boy.

Word Quest Generator-A portion of the curiosity around puzzles will build their own for many kids. On the Internet, many word search creators will find or build their own from scratch. It's not a difficult operation, so have fun together with your child and make something. The creation process is sometimes as enjoyable as the puzzle itself.

Term Unscrambles-It 's the same the world over, regardless of what some people call this game. You have several letters that can be used to render other sentences. This forces the child to think beyond the box and build options that might not be easily visible on the table. The more letters you have, the more anagrams possible, and the more you will need to help younger kids think about possible alternatives.

Make Your Own Jigsaw Puzzle-There's hardly a kid out there that doesn't enjoy jigsaw puzzles being solved. Then why not have your kid make his or her puzzle jigsaw? You need to take a photo and make him or she cut puzzle pieces out of it (that is safe to cut up). The younger they are, the bigger the bits need to be. To make it easy to bring things back together, first stick the illustration onto a thin cardboard or poster board. That way, every piece is going to be solid and remain together.

Scrabble-Oh, that's it, yes. Fucking scrabble. These days, there are Scrabble variations for children of any generation, and this is one of the easiest opportunities to get the kids to learn words, grammar, and problem-solving.

Up there, there are so many games that are suitable for children to enjoy. There are only five wonderful puzzles that children enjoy.

CHAPTER 9: CAN EDUCATIONAL GAMES FOR KIDS HELP?

Educational games are for girls everywhere. They range from real-life board games to online games that are free. They can be rented or easily downloaded as well. There are toy sections in department stores where hundreds of games can be found. Although not all of them can be educational, finding scrabble and chess games is not easy. Famous websites such as Facebook and Yahoo supply a number of games, most of which are child-friendly.

Parents should also create games for their children. If parents are creative enough to add a little twist on it, it will become a worthwhile job to solve a simple math problem. It takes ingenuity and resourcefulness to do this.

Word games are the most common ones. Some of the more popular forms are Hangman, Boggle, crossword puzzles, Scribbage, and the most notorious scrabble. These games allow more vocabulary for kids to understand. Difficult terms will cause them to scan the dictionary and discover the meanings in the process. By having two or more kids play against each other, Boggle, Scrabble, and Cribbage will provide a competitive advantage for kids. Although crosswords have been developed to please adults, basic crosswords are found in some instructional books for children. Parents, including their teenagers, may also do crosswords of their own or better when referring to their favorite newspaper crosswords.

There are also board games that can assist the learning of a boy. Not only are flashcard games confined to simple mathematical operations. These children's interactive games will also extend their concept of wildlife, nature, and sports ideal for toddlers.

Children's educational games can also be in the form of computer games. These games have undergone an evolution from basic to strictly educational entertainment. Some researchers have also indicated that some video and computer games can be used for serious studying in a classroom environment. This is parallel to the advances achieved by the world with the advancement of digital technologies. Strategic thought is promoted by some of these sports, while some rely on direct instruction.

The online world is also flooded with interactive games for children. Only one Google search can give you thousands of website choices that can be detailed, fun, and useful for learning. The top-of-the-list websites include learninggamesforkids.com and playkidsgames.com. In terms of music and the arts, fitness, and science, the former has several games. You will create a classroom at playkidsgames.com to get the kids together to play in an online class atmosphere.

Nevertheless, not all individuals identify with this scheme. Questions about the efficacy of technology and the Internet for students are being posed. The key concern is that web content is not monitored, but the Internet can be potentially dangerous without adequate guidance.

Parents and teachers should make sure that the kids stick to the route they should follow to school. There are instructional games for children as substitutes, but their utmost focus must be to protect the kids.

Evaluate Reading Programs for Kids

One of the biggest obstacles for parents has been finding the right curriculum for children with learning disabilities. After all, parents must help their children face life's realistic difficulties and find their place in society. With software tools, one-on-one training, and personalized instructional systems, professionals have defined and invented new teaching strategies, many of which are considered to help these children deal successfully with their vocabulary and word identification issues.

Individual-Specific Reading Programs

A wide variety of services have been planned and implemented to support children who suffer from writing and reading due to dyslexia and associated difficulties. Different specialized services, such as comprehension, phonemic knowledge, grammar, understanding, writing, and spelling, concentrate on various talents. You would be able to make the right decision for your child, depending on your child's needs and with clinical assistance.

Development of Word Recognition Skills

To support children with dyslexia and other learning disorders, several word processing systems and phonics with detailed guidelines have been developed and integrated by professionals. Specific emphasis on vocabulary, phonics, and word recognition helps develop and enable essential awareness to recognize text material when reading, enhancing the kid's fluency. Other elements, such as history, mathematics, music, science, etc., require terms unique to the content and should also be taken care of.

Dr. Swanson, Ph. D, helps you determine the right reading services and alleviate the uncertainty. The University of California's leading psychology professor has examined and established the most successful approach to developing children's recognition capabilities with LD by segmentation, sequencing, and specialized organizers.

Many of the most powerful technologies for teachers to adapt to help children understand vocabulary include:

• Phonics

- Decoding

- Word attack skills

- Phonemic awareness

Evaluating the Reading Program for your Kid

You should go through the following instructions for the correct understanding of the best training methods to test the child's reading program:

A comprehensive plan and list of personalized programs to learn specialized courses or organizations offer your child's unique talents. You will have to press for the same if they do not provide it. Clarify all your questions about and program on offer (especially those that are acceptable for your child) and ask if your child will benefit from it.

Be mindful of the full teaching model and techniques tailored for your child, as some literature programs define the instructor's method.

Have frequent sessions with your child's teacher to see how anything goes according to the timeline and if your child does or reacts.

While all of these need to be achieved with caution, doing your part, and providing your full child protection at home is equally important, Remember-it is the awareness and inspiration that will enable the child to face life with confidence.

HOW IS AN IPAD USEFUL FOR KIDS?

Many parents use numerous iPad apps to make their children's education more exciting and engaging. Math Bingo, Math Screen, Fish Academy, Create a Zoo, and 123 World Geography are some of the great iPad games for kids. Among both parents and children, these are becoming a significant hit. The best iPad educational applications for children built into the iPad are specially developed by professionals and are good tools that can help develop children's understanding.

One of the most useful children's iPad apps is Math Bingo. This app involves learning mathematical principles through a puzzle, such as multiplication, addition, division, and subtraction. The game is easy, and it helps to develop the mental capacity of children. For those children who are poor in mathematics, this software is extremely helpful. They will learn the basic principles of mathematics and improve their general math skills. Not only does an iPad make your life easy, but it also supports your children in kindergarten.

Children will also learn more about history, geometry, physics, geography, and more and help with math problems and improve concepts. Applications such as Word Magic and Bookworm can encourage children to learn new words. These are the best iPad software that supports kids at home while away from their school teachers' supervision. Besides, there are other children's iPad apps that help to learn grammar and grow their vocabulary. Therefore, for kids who can use it and get many perks, iPad applications are helpful. This software goes a long way towards helping children learn more about and build awareness about current events unfolding worldwide.

Kids still require some amusement and fun apart from school. Since getting back from training, they get sleepy. Therefore, to do their homework and assignments properly, it is necessary to recharge and refresh them. Your iPad will be seen here, too. It has several games specifically for children such as Pac Man, Chuzzle, Lunchbox, Spiderly and Monkey Playgroup, etc. 'Spiderly' is another excellent iPad title. This title is a fascinating idea that needs a sharp focus. In specific, this game promises fun and also develops the mental capacity of the boy. Therefore, one should inspire kids to play games like these.

Kids have some clever applications for the iPad. When they play, it encourages them to think. Not all children like to sit down and try to solve problems with books. They continue to learn to be more fascinating and special. Via its special applications, iPads offer just the right learning experience by integrating play into it. You will look back and watch their grades rise as your children love playing sports.

"You would be able to read this piece as you drive your car to go to a family picnic, and one of the kids pipes up every five minutes to inquire," Are we there yet? "because I bring you six applications that are made only for kids. So, all you have to do is hand the iPhone to them and let them play a game or watch some fun and informative content. I sorted the applications according to the age of the girls. Here are the six best children's iPhone applications:

Apps for Toddlers

Where is Gumbo is an animation game in which it is intended that children can search and peer behind multiple items in a scene to search for the cat, Gumbo, the dog? Gumbo and other pets are put randomly behind the assorted items, so any time they play it, the child should expect an original game.

For artistically inclined children, Colorama is a game, and at this age, which kid is not artistically inclined?! The game comes with more than 50 illustrations, as they wish, colored by the boy. The sketches can also be saved and emailed after they have finished the coloring, from where you can print them out as well. It would be a wonderful opportunity for any child to see a picture built by themselves over their crib!

For younger ones, ages 4-10

Scoops This game includes building a tower out of ice-cream scoops. The imaginative way in which it uses the accelerometer is what makes this game unique. In order to get the scoop right on top of the house, you should tilt the screen.

Brain Toot is a game that includes basic, entertainingly presented mathematical issues. And no, it doesn't consist exclusively of what is 4 + 3? It is more intriguing and demanding than that. Full activity for the little grey cells, as indicated by its name.

For Older Children

Scrabble is one of the world's most popular word games.

From Zynga's original business, this game is a safe way to keep kids occupied and learn new vocabulary. The game is multi-player, meaning two or three children will use the same computer to play it.

Another classic game that has been downloaded over 150 million times, not only as an iPhone version, is Bejeweled. It consists of a board full of gems that need to be swapped to ensure that there are three or more gems of the same form and color in a row, horizontally or vertically. On the iPhone, the game becomes all the more fun with its touch interface.

CHAPTER 10: WHEN HANDWRITING IS AN ISSUE

For children with a sensory processing disorder, learning handwriting abilities, and handwriting for longer than a minute or two is always challenging. Handwriting has many difficulties, including recall, language processing, balance, muscle tone, body perception, tactile problems, etc. If your child has trouble handwriting, ask your school to get a sensory smart occupational therapist (OT) to diagnose your child with experience in handwriting problems. You and the OT will work on them and help your son or daughter learn proper penmanship until you can recognize which challenges are at play. Meanwhile, he will need to learn to improve the composing element of writing as your child grows older.

It will profoundly assist a child who has the inconsistent ability as you distinguish composing from handwriting. There is nothing more annoying than understanding what you want to say and not getting it written with a pencil on paper unless it doesn't know what to say and has booting issues with handwriting! Too frequently, we mix the many distinct elements of writing and editing that can scare and confuse an inexperienced writer, especially one who grapples with physical handwriting difficulty. After all, years ago, managers frequently composed their letters through dictation, and secretaries used a Dictaphone or Gregg shorthand to document what they were saying. There is no need to demand that the two talents, handwriting and composing, often go "hand in hand." Via dictating them to a secretary, bestselling romance author Barbara

Cartland wrote all her novels.

Some Few tips to help your child handle the composition part of writing without being irritated and nervous.

1. For composing, utilizing technologies. Provide assistive devices such as a keyboard, iPad, or dictation apps such as Dragon Spontaneously Speaking (R), or make your child's school provide

it. Training such software requires time, and it could be difficult for certain children to understand, so make your child test the curriculum before committing to using it.

2. Using dictation that is old-fashioned. As you write it, your child will dictate his book report to you or someone else. On your smartphone, a low-tech tape recorder, or other technology, you can record it, then play it back and write it out for her.

3. For a few minutes for rehearsal, make her compose openly. Using handwriting or a keyboard, please make your child sit and write whatever she likes. This is going to foster self-expression. Praise her; please don't make corrections for the initiative. Let her get used to the fact that she can compose her thoughts and "write." Have a quick one if she's stumped on a topic, and praise her for writing something on the subject. There are books loaded with lists of prompts for writing tailored to particular ages of age. If she is nervous, start writing for as little as one minute, small, free.

4. Encourage the drafting of short-form letters and messages. A child who composes letters to his cousins and scribbles on the family blackboard amusing little notes to Mom and Dad will have an

easier time approaching a broader writing challenge than a child who barely practices writing to express himself.

5. Reflect on emotions and how they are connected first. For certain girls, visual mapping using bubbles, or Inspiration apps, which enables you to do this conveniently on a screen, works well. Before beginning the writing process, other kids need to talk them over with a parent or teacher.

6. Next, emphasis on arranging thoughts and sentences. Children with sensory difficulties also have a very difficult time arranging time, belongings, and feelings. They do not know that there should be a beginning, middle, and end of a report or letter or that a sentence has such features that make it a full sentence. Before looking at the fundamentals of grammar, punctuation, and capitalization, focus on certain structure aspects. It will help your child understand the writing art as composing and taking away the burden to recall all those graphic bits (such as capital versus small letters) involved in writing on paper or the computer.

7. Let her pick the font and its size while working with a machine. Crazy as it may seem, whether

they may select a font visually appealing to them, certain children with sensory disabilities can have a better time writing on a computer screen, using a keyboard. Before printing it, you can still change it later. The style will need to be very small or very big for your child with vision disabilities or feel that the letters are easier for her to read in a font you feel too "out there." Consider changing the computer screen to minimize glare and have more or less contrast, based on what makes your child most relaxed. Our emphasis on handwriting and its sensory challenges is that computer screens and computers, too, will forget sensory issues!

8. Teach your kid that it's possible to delete later. Most of us edit to a degree while we compose, but a nervous child will immediately get swept up in "making it right." The writing method will confuse children with autism, who are rigid learners and appear to have anxiety. Teach your child that even the best authors go through several drafts of what they write, and the first level of editing is to reflect on the thoughts and how they are presented. Yes, if she finds that she failed to capitalize the first letter or misspelled a word at the beginning of a sentence, she might fix that, but it's not what she should be looking for until she's made sure the ideas convey

the way she wants them to sound. It can be immensely useful to read the composition out loud.

9. Act independently on pronunciation. When handwriting is not involved, you will find your child's spelling is better. He frees himself up to pay more attention to his spelling by not stressing on the handwriting aspect, playing with making him dictate how to spell the keyboard words. By keyboarding them or spelling them aloud, make him go over spelling words-perhaps while going in a circle, turning on an office chair or Dizzy Disc Jr.(R), or jumping on a mini-trampoline. Some children find that understanding the roots of words is very good for spelling.

10. When you break the mission down, keep the mood optimistic. Whenever your child shows worry over a major assignment, be optimistic, and break down the big assignment into smaller assignments. A great journey begins, as they say, with one small step!

Online Reading for Kids

The internet has become vital for students of all ages worldwide, but it has become increasingly important in reading for young children. Children these days love to play games on the internet at a very early age, as computers and the internet are readily available at home and school. Many internet businesses have converted their online games into children's learning modules, especially those that teach reading and reading awareness. Many of the reading websites providing children online reading are free of charge, but some of the more active offer subscriptions and numerous classes for children to teach reading.

One of the strongest aspects of children's online reading is that websites and electronic books are offered in thousands of languages, ensuring that zero language gaps will lead them to fail when learning to read. Without any issues, most blogs and books may either be translated into the child's language or bought in the child's language. The ease of access is another advantage of children's online reading, where there are no actual paperback or hardback books readily accessible for the child to use. Physical books are not often available, but iPods, iPads, iPhones, BlackBerry devices, Android phones, notebooks, and eBook readers can be used for online reading. This means that no matter where they are placed, your child will be able to read online.

Image books that are accessible on the internet also provide the online reading for children. Picture books will let kids know their names for dogs, cats, donkeys, and other species through word association. These days, the easiest way to learn a foreign language is to use the internet, regardless of your age. For children who are learning a second language, the same may be said. Today, hundreds of blogs and eBooks for language instruction are on the internet that will instruct children who use online literacy to aid education.

Digitized audiobooks under the age of two will have the text read aloud if the child is a very early learner as the child stares at the text on the computer. When looking at the letter, hearing the letter read aloud to them will help young kids learn the sounds of words and what they look like. Online reading for children is also helped by cultural curiosity because it encourages them to read works from other countries worldwide. Kids would be able to see what literature from other cultures is like and why they use photographs. Children's online literacy has been an ideal way to show young people the ins and outs of literacy. This involves studying bigger phrases in the United States, studying to understand reading, learning to fit pictures with phrases, and learning correct grammar.

CHAPTER 11: QUESTIONS SECTION

Q1: There will be twice as many dogs as cats on our streets. How are we expected to write this as an equation?
- Let D = number of dogs
- Let C = number of cats

Now ... is that: $2D = C$

or should it be: $D = 2C$

Think carefully now!

Q2: Sam's got 2 dollars far below Alex. How are we expected to write this as an equation?
- Let S = dollars Sam has
- Let A = dollars Alex has

Now ... is that: $S - 2 = A$

or should it be: $S = A - 2$

or should it be: $S = 2 - A$

Q3: Sam and Alex play Tennis. Sam played 4 more games on the weekend than Alex did, and they played 12 games together. How many games were played by Alex?

Transform it into Algebra in English:
Lyrics:

For how many games Sam played, use S
Use A to mean how many games Alex has
played
Sam played four more games than Alex,
we remember, so: S = A + 4
And we know that they played 12 games
together: S + A = 12.
We are asked how many games Alex has
been playing: **A**

Q4. The girls had to sell tickets for their
games for three weeks. They sold 75
tickets within the first week. They sold 108
tickets in the second week and sold 210
tickets in the third week. How do they sell
tickets at all?

Tickets sold within the first week = 75
Tickets sold during the second week =
108
Third-week tickets sold = 210
Number of tickets sold in total = 75 + 108
+ 210 = 393

5. Q5. On Wednesday, Mr. Bose spent
$450 on fuel. On Thursday, he spent $125
more than that. How much did he spend
over those two days on petrol?

You have to solve this issue in two phases.
Step 1: The cost for petrol on Thursday
125 +450 = 575

 Step 2: Both days the money spent on petrol on
575 +450 = 1025

Examples of word problems on addition and subtraction:
1. What is the sum of 3127, 4373, and 4191?

Q6. Subtract 4358 from the sum of 1324 and 5632.

Q7. Find the number, which is
(i) 1240 greater than in 3267.
(ii) 1353 smaller than in 5292.

Q8. The town's population is 16732. If there are 9569 males, find the town's number of females.
Q9. One factory has 35,675 employees. In the first shift, there are 10,000, seven hundred and fifty workers; in the second shift, 12,650 workers; and in the third

shift, the remainder. For the third shift, how many people are there?

Word multiplication problems are solved step by step here for fourth-grade students.

Multiplication Involving Problem Sums:

1. There are 24 directories with 56 sheets of paper inside each. In total, how many sheets of paper are there?

Solution:

Since multiplication is repeated addition, to get the answer, we can multiply 56 and 24.

[More than one calculation is needed sometimes.]

There are 1344 sheets overall, thus.

2. In a carton, 24 packets of biscuits are kept. Every packet has 12 biscuits in it. How many biscuits will they pack in 45 cartons?

In one carton, we can pack 24 × 12 biscuits.

In 45 cartons, we can pack 24 × 12 × 45 biscuits.

24 × 12 = 288

→ 40 + 5
Multiply by 5
Multiply by 40

Therefore, altogether there are 12,960 biscuits.

Find the following examples of word multiplication problems:

Q1. It costs $67 for a book. How much will 102 books like this be paid for?

Q2. It costs $215 for a bicycle. How much will 87 such bicycles be paid for?

Q3. A man's monthly salary is $2,625. What is his annual pay revenue?

Q4. A chair costs 452 dollars, and a table costs 1750 dollars. What is the expense of fifteen chairs and 30 tables?

CHAPTER 11: ANSWERS SECTION

Q1 ANSWER: The response is D = 2C
(2D= C is a common error, as "twice ...
dogs ... cats" is written in the question)
Q2 ANSWER: S = A − 2 is the right
answer
(S-2 = A is a common error, as "Sam ... 2
less ... Alex" is written in the question)

Solve Q3:
Begin with: S + A = 12
S = A + 4, so we will be able to
Substitute S for "A + 4":(A + 4) + A = 12
Simplify the following:2A + 4 = 12
2A = 12 − 4 Deduct 4 from both sides:
Simplification:2A = 8
Divide by 2: A = 4 on both sides
Which means Alex played 4 tennis games.

Check: Sam has played four games more than Alex, so Sam has played eight. They played 4 + 8 = 12 games together. Yes! Yes!

Q4 Answer: In all, 393 tickets were sold.
Solution Q5:
The figures are grouped and added in columns.

Q4 Answer: In all, 393 tickets were sold.
Solution Q5:
The figures are grouped and added in columns.

Q4 Answer: 393 tickets were sold in all.
The figures are grouped and added in columns.

Q4 Answer: In all, 393 tickets were sold.
Solution Q5:

The figures are grouped and added in columns.
1. Ones are added: 7 + 1 + 3 = 11 = 1 Ten + 1 one
2. Tens are added: 9 + 1 + 7 + 2
= 19 tens
= 1 hundred + 9 ten
3. Hundreds are added: 1 + 1+ 1 + 3= 6 Hundred
4. Thousands are added: 4 + 3 + 4 = 11 Thousand

Therefore, sum =11,691
2. What is the difference between 1298 and 3867?
Solution:

Numbers are arranged in columns and subtracted:
1. Ones are subtracted: 7 < 8
1 is borrowed from 6 ten. So, 1 T or 10 + 7 = 17,
17 - 8 = 9
2. Tens are subtracted; 5 T < 9 T,
So, 10 T or 1 H is borrowed from 8 H, 1 H
= 10 T + 5 T = 15T
15T - 9T = 6

3. Hundreds are simply subtracted 7 H – 2 H = 5 H

4. Thousands are also subtracted 3 Th – 1 Th = 2 Th

Therefore, difference = 2569
Q6 Solution:

 Sum of 5632 and 1324

Difference of 6956 and 4358

  1. 6 < 8, 1 T or 10 ones are borrowed

1 T or 10+ 6 =16, 16 - 8 = 8

2. 4 T < 5 T, 1 H or 10 T is borrowed

10 T + 4 T = 14 T, 14 T – 5 T = 9 T

3. 8 H – 3 H = 5 H

4. 6 Th – 4 Th = 2 Th

Therefore, 2598 is the answer.
Q7 Solution:

 (i) The number is 1240 more than 3267

Therefore, the number = 3267 + 1240 or = 4507

 (ii) The number is 1353, less than 5292

Therefore, the number

= 5292 – 1353 or

= 3939

Q8 Solution:

The population of the town
Number of males
Therefore, the number of females

= 1 6 7 3 2
=   - 9 5 6 9
= 7 1 6 3

Q9 Solution:

The total numbers of workers coming in the first and second shift
=12650 +10750 = 23400

The number of workers coming in the third shift = 35675 - 23400 = 12275
Problem Sums Involving Multiplication Answers

Q1 Solution:

Here is the Cost of one book = 6 7 6 7

The total Number of books = 1 0
2   × 1 0 2

The cost of 102 books = 102 ×b 67 1 3 4

        = $ 6834   + 6 7 0 0

6 8 3 4

**Therefore, the cost of 102 books =
$ 6834**

Q2 Solution:
This is the cost of one bicycle = 2 1 5 2 1 5

Here Number of bicycle = 8
7   × 8 7

The cost of 87 bicycles = 87 × 2 1 5 1 5 0
5

= $ 6834   + 1 7 2 0 0

1 8 7 0 5

**Therefore, cost of 87 bicycles =
$ 18705**
Q3 Solution:
The Monthly income = 2,625 2 6 2 5

The Annual income =12 × 2,625   × 1
2

= $31,500 5 2 5 0

$$+ 2\,6\,2\,5\,0$$

$$\underline{3\,1\,5\,0\,0}$$

Therefore, annual income = $ 31,500

Q4 Solution:

(i) Cost of one chair = $ 452

Cost of 15 chairs = $ 452 × 15

= 6,780

(ii) Cost of one table = 1,750

Cost of 30 tables = 30 × 1,750

= 52,500

15 chairs and 30 tables = 6,780 + 52,50000

Lightning Source UK Ltd.
Milton Keynes UK
UKHW020636200421
382299UK00011B/698